MATH IN PRACTICE

TEACHING Kindergarten Math

MATH IN PRACTICE

TEACHING

Kindergarten Math

Marcy Myers • Susan O'Connell • John SanGiovanni

HEINEMANN
Portsmouth, NH

Heinemann
361 Hanover Street
Portsmouth, NH 03801–3912
www.heinemann.com

Offices and agents throughout the world

Cataloging-in-Publication Data is on file with the Library of Congress.
ISBN: 978-0-325-07824-3

Editors: Katherine Bryant and Sue Paro
Production: Victoria Merecki
Typesetter: Publishers' Design & Production Services, Inc.
Cover and interior designs: Suzanne Heiser
Manufacturing: Steve Bernier

Printed in the United States of America on acid-free paper
20 19 18 VP 3 4 5

To my four Ms—Mike, Maggie, Mitch, and Max
MM

To Kellen with love
SO

To Oscar and Deryn
JS

Contents

Acknowledgments

This project was a collaborative effort and could not have been done without the contributions of many people. A special thanks to the following principals and teachers who welcomed us into their schools and classrooms to work with students, listen to their mathematical thinking, and watch them at work doing mathematics: Darryl Robbins, Betsy Cunningham, Justin Watts, Sarah Parsons, Brittany Collett, Nicole Mort, Dianne Russell, Samantha Havighurst, Kaymi Plank, Janet Shearer, Deb Zukowski, Ashley Brewer, Mel Rhoads, Suzanne Peters, Jen Purper, Karlye Cowherd, Donna Gury, Kerry Rosier, and Nicole Trotz.

Thanks to Kim Kucharski and Donna Gury for sharing classroom ideas and teaching strategies that will undoubtedly be valued by many teachers. Thanks to Stephanie Ross for the wonderful photos.

We are grateful for the assistance of Sue Paro and Katherine Bryant in reading and rereading our ideas and providing helpful suggestions.

We have enjoyed the opportunity to work with an amazing writing team for the Math in Practice series and have enjoyed our collaborations and many, many meetings with Laura, Allison, Cheryl, Kay, and Joan.

MM, SO, JS

I am grateful for the support of Laura Hunovice, Carey Sneska, Donn Hicks, Beth Kobett, and Dr. Francis "Skip" Fennell. Thanks to my family for their constant support and encouragement throughout this project.

MM

Introduction

This book is about the teaching of mathematics to students in kindergarten. It does not focus solely on the content taught in kindergarten, because we recognize that math is about more than content. Our students are challenged to be able to use their math skills and knowledge, to be mathematical thinkers. It is not a book providing scripted lessons, because we recognize that the teaching of mathematics is not about reading scripts, but about listening to our students, making decisions, and having options for ways to support their math learning. It is not a book of activities, because we know that you can find activities everywhere, but rather a book filled with strong lesson ideas that are based on research-informed instructional strategies that can be modified to meet your students' needs. It is about providing you with options that allow you to choose effective teaching strategies and identify tasks that link to the standards expected of your kindergarten students.

This book is more about the teaching of students than the teaching of content. It addresses ways to engage students, help them visualize math ideas, support them as they solve problems, help them make connections and develop insights, build their confidence in math, and learn to love mathematics. Students who think mathematically, reason and solve problems, understand the way numbers work, and have a positive disposition about math have truly met today's expectations. This book is about helping each of our students develop those skills.

A Focus on Teaching and Learning

We have known for a number of years that telling students how to do math procedures, asking them to memorize facts and formulas, and focusing on answers as right or wrong have not resulted in students who are skilled and confident in their math abilities. Today's standards have pointed us toward the adoption of teaching practices that value understanding and application, but it has been difficult to know just how to implement that within our classrooms.

Today's classrooms are standards-based classrooms. The lesson ideas and instructional practices throughout this book align with the National Council of Teachers of Mathematics (NCTM) Standards (2000), as well as the Common Core State Standards (CCSS). The content progressions in this book align closely with the progressions found in the Common Core standards. The instructional strategies are not an arbitrary list of strategies, but rather a compilation of consistently recommended approaches in line with the NCTM Process Standards, the Common Core Standards for Mathematical Practice, and the Mathematics Teaching Practices highlighted in *Principles to Actions* (NCTM, 2014). They are informed by research and consistent with our current standards.

About the Math in Practice Series

This book is a part of the Math in Practice series, a comprehensive resource designed to identify critical math content, share essential instructional strategies, and provide resources to support teachers, administrators, and school communities as they rethink the teaching of K–5 mathematics. The series consists of eight books.

- *Math in Practice: A Guide for Teachers* outlines research-informed instructional practices that enhance math teaching and learning. It offers a rationale for changing the way we teach math to K–5 students and explores simple ways we can adjust our teaching to improve students' learning. A variety of highly effective instructional strategies are introduced, discussed, and illustrated throughout the book.
- Six grade-specific books, like this one, focus on the specific mathematics taught at each grade level, K–5. The instructional strategies introduced in the *Guide for Teachers* are shown throughout the lessons in the grade-specific books. These books include discussions about the math being taught, sample lessons that blend content and practice, a wealth of instructional tasks, practical teaching tips, and ready-to-use classroom resources.
- *Math in Practice: A Guide for Administrators* provides school and district administrators with an overview of math teaching and discusses what to look for in math classrooms. The book includes tips for guiding the math program and ideas for supporting teachers as they work to improve their teaching.

A variety of instructional strategies are introduced in *Math in Practice: A Guide for Teachers*. You will notice these instructional strategies illustrated throughout the lessons in this book. These strategies build deeper math understanding, promote mathematical reasoning and problem solving, strengthen math communication skills, and provide ongoing assessment information regardless of the math content being addressed. As you explore this grade-level guide, you will see how these instructional strategies enhance teaching and empower your students. The following instructional strategies are integrated into the lesson ideas you will explore in each module:

- problem solving integrated with content and skills
  - posing problems to provide a context for math skills and to challenge students to apply those skills

- use of representations (models) to build math understanding
  - encouraging the use of varied models, including manipulatives, drawings, and graphic representations, to explore and refine math understanding
- posing in-depth questions to prompt thinking
  - asking students to describe math concepts, explain how they performed math tasks, justify solutions, and predict outcomes
- ongoing opportunities for talk and writing about math ideas
  - challenging students to show and explain their thinking through both talk and writing
- connections to real-world situations
  - using real-world connections to help students make sense of mathematics and its abstract numbers and symbols
- formative assessment to drive instruction
  - posing ongoing tasks to determine students' level of understanding and then using the data to adjust instruction to support students' learning
- differentiation to meet the needs of varied learners
  - modifying our tasks and our delivery of instruction to ensure that all students have opportunities to learn mathematics
- focus on the development of mathematical practices
  - going beyond memorizing content to help our students think like mathematicians and do the work of mathematics.

About This Book

This book is organized into modules. Each module addresses an important math topic for kindergarten students. It can be read from start to finish to provide an in-depth look at kindergarten mathematics or it can be read by module to provide tips, insights, and strategies as you explore a certain math topic with your students. Some important points about the organization of this book:

- **The modules do not represent an order in which to teach the content.** They are simply a way to organize the math ideas so you are better able to access what you need at any point in time. Each module provides you with an overview of a math topic, as well as ideas for instruction and assessment related to the topic. This book is organized in modules to allow you to find the topic easily and see the development of that topic from simple to more complex ideas. Your state or school district likely has a pacing schedule or a sequence through which they suggest you teach the grade-level standards. This book is organized so you can access the information you need to enhance the teaching of any math topic.
- **This book does not recommend teaching by topic.** We do not suggest that you teach by modules. In fact, we highly discourage it! Math is not a set of isolated

skills and should not be taught in that way. We strongly recommend revisiting topics throughout the year at different levels of complexity and connecting each new concept to a previously learned concept. Teaching all measurement skills at one point during the year, for instance, and then moving on, does not give students the ongoing exposure they need to master and retain those skills. A measurement module gives you a wealth of ideas about teaching measurement, but your district pacing guide will help you identify when specific skills are introduced and revisited. It is likely you will be back in the measurement module at various points throughout the year as your students revisit skills or transition to more complex ones.

- **This book is about choice.** We encourage you to select lesson ideas and build your lessons from them. Teaching well is not about using every activity in the book. It is about finding those that address the needs of your students. For that reason, each module offers a wealth of additional ideas to get you thinking about options for teaching, reteaching, or enrichment.
- **This book does not specify whether lessons should be done with all of your students or just some of your students.** Although in one classroom all students may benefit from an activity, in another classroom it may be the perfect choice for a small group of students who need additional exposure to a skill. Consider each lesson idea and how it might fit best with your students.

This book is a book for teachers, not students. It is organized to allow you to find topics, read about the critical math concepts, reflect on instructional activities, and use this information to plan and facilitate the best lessons for your students. It is filled with lesson seeds rather than a set of scripted lessons. With your consideration, reflection, and modifications, these lesson seeds will grow into classroom lessons that engage and excite students and stimulate math learning.

Module Organization

Each module focuses on a key math concept at this grade level. It highlights the important math ideas and is filled with lesson ideas, practice tasks, assessment options, center tasks, discussion starters, and key vocabulary. Understanding the structure and key elements of each section of the module will help you better utilize the ideas and activities presented.

About the Math

The About the Math section describes the important math concepts being addressed throughout the module. Rather than simply listing math skills, this section provides a glimpse at the big ideas that underlie the skills and helps you identify key teaching points for the topics discussed within the module. It provides a glimpse into teaching the understanding of many skills that were previously taught in rote ways, offering you a better idea of how to build that understanding in your students.

This section includes a brief table to show the progressions for the skill, indicating the related skills and concepts at the grade level before and after. This table gives you a quick glimpse of what is introduced at this level and what is a continuation from the previous grade level. When appropriate, connections are made between the skills and concepts in this module and related skills and concepts in other modules. Math is not a series of isolated topics. The goal is to reflect on how each topic is connected to others to present math ideas in a connected way.

This section also includes:

- a brief overview of the mathematics related to the topic
- a discussion of the big ideas that underlie the math skills/concepts
- what is new this year and what is continued/expanded from previous years
- progressions for related skills in the grades before and after
- connection to the content standards for this grade level (We use the Common Core standards for reference, although the mathematical content is relevant to whichever standards you use.)
- suggestions/recommendations for teaching strategies that build these skills
- a list of *I Can* statements to highlight student learning goals in student-friendly language.

Ideas for Instruction and Assessment

This section presents a variety of lesson ideas, practice tasks, and assessment options. You will not find scripted lessons in this book, because lesson planning should be specific to your students' needs and abilities. What you will find are lesson ideas that get you thinking about what to teach and ways to teach it.

This section presents lesson ideas and assessment options related to the key ideas highlighted in the About the Math section. You may choose to implement these very similarly to the way in which they are described or to modify them depending on the needs of your students.

Our goal is twofold: (1) to provide you with a wealth of classroom ideas so you can select learning experiences that work for your students and (2) to elaborate on some tasks to allow you to experience a teacher's thought process as he or she works with students to build their understanding and math confidence. We want you to have lots of classroom ideas, as well as insights into ways you can enhance each activity for optimal student learning. To achieve both goals, you will notice three types of lesson structures:

1. **Lesson Ideas**—These appear early in the Ideas for Instruction and Assessment section and contain a description of the lesson, including ideas for teacher questioning that promotes student understanding. Rather than simply saying "Probe for understand about . . ." or "Ask students to discuss . . .," a series of questions

appear to allow you to hear the teacher's questioning process and imagine the student responses the questions might elicit.

2. **Thinking Through a Lesson**—In each module, one lesson idea has been elaborated on to become a reflective lesson. These lesson ideas contain the lesson description and teacher questioning, just like those above, but also include more detailed teacher notes that appear at various points throughout the lesson. These notes highlight teacher observations and suggestions as the lesson unfolds. They provide you with the behind-the-scenes thinking that guides us as we observe our students, make decisions, modify tasks, and provide needed support. These lessons also include references to the Standards for Mathematical Practice, highlighting opportunities to foster students' use and understanding of these key ways of thinking about and doing mathematics.
3. **Additional Ideas for Support and Practice**—These additional ideas offer ways to develop skills and understanding for students who may need more exposure or ways to engage students in meaningful practice. They might become a whole-class lesson if additional instruction is needed, a small-group lesson for some students who would benefit from another exposure to the topic, an extension for those who need to be challenged, a center task for partner practice, or a way for students to independently practice the skill. They are meant to give you additional options to extend students' learning.

ICONS

You will notice specific icons as you are perusing lesson ideas. These icons connect us back to *Math in Practice: A Guide for Teachers* and the key instructional strategies that were highlighted. At any point, you can refer back to that book for more details on the strategies. These icons include:

K.CC.3 **Standards**—When this appears in the About the Math section of a module, it indicates the specific content standards addressed within that module. When it appears in a Thinking Through a Lesson feature, it indicates the Standards for Mathematical Practice that appear within that lesson.

SHOW IT! **Show It**—This indicates that skills and concepts are made visual through concrete objects or drawings.

Talk/Write—These discussions starters or writing prompts can be used to get students talking and writing about the key ideas related to the topic.

Literature—These lesson ideas incorporate the use of children's books to provide a context for the math skills and concepts being taught.

Differentiation—These ideas help you identify ways to differentiate instruction, offering ideas for modifying tasks or modifying your delivery of instruction for either students who are struggling or those whose learning should be extended.

Centers—These ideas offer a variety of ways for students to practice skills, alone or with partners, including ideas for math centers.

"I CAN" STATEMENTS

At the start of each module, you will notice "I can" statements that illustrate the instructional goals for the module from a student perspective. The activities and assessments throughout the module serve to guide students to these skills or understandings.

FORMATIVE ASSESSMENT

An essential component of effective math teaching is ongoing assessment to determine what our students know and what they may be struggling to understand. Throughout the book, you will notice formative assessment options that allow you to check for student understanding related to the math that is highlighted in each module. You may see samples of student work to help you visualize student responses, as well as suggestions for dealing with common errors. Ongoing assessment allows us to modify our instruction based on what our students know, or don't know, and keeps us focused on student growth.

VOCABULARY

A list of related math vocabulary words appears in each module. These words, although not an exhaustive list, represent some of the most critical terms related to the module's content. Discussing the meanings of these words, posting the words on a class Math Talk chart or word wall, and expecting students to use these words as they explore the math ideas deepens their understanding of the content and supports their ability to communicate with precision. Ideas for ways to develop these words are shared in Chapter 4 of *Math in Practice: A Guide for Teachers*.

Online Resources

Although you will find many resources throughout the pages of this book, additional resources are available on the accompanying website: http://hein.pub/MathinPractice. You will notice thumbnails throughout the modules that give you a quick glimpse of some of the accompanying materials. You will find:

- templates for lesson recording sheets or practice tasks
- resource materials like ten frames, number cards, hundred charts, or spinners
- ready-to-go center activities
- problem-solving cards related to the module topic for additional problem-solving practice.

These materials make your implementation of the lessons and activities easy and manageable. You will notice that many of the materials are in Microsoft Word rather than a PDF format to allow you to easily modify them to meet the needs of your students. Change the names to be names of your students, modify the data to make them simpler or more complex, or delete parts to shorten the task.

> To access the online resources, visit http://hein.pub/MathinPractice. Enter your email address and password (or click "Create a New Acount" to set up an account). Once you have logged in, enter keycode MIPGK and click "Register."

How to Use This Book

As you read this book, consider the many ways you can use it to enhance your math teaching and support your students' learning. Read it to:

- better understand and gain insights about the math content at this grade level
- gain a deeper understanding of what comes before and after what you teach, and how our students' math understanding builds in a progression over several years
- explore the level of questioning that builds deep math understanding
- discover ways to make math concepts visible to all students through opportunities to investigate math in hands-on and visual ways
- reflect on ways to lead student investigations rather than telling students how to do math
- find formative assessment ideas that allow you to identify what students know and reflect on sample student responses to find strategies to address what they struggle to understand
- gather a wealth of ready-to-use center and practice ideas that engage students in doing meaningful mathematics
- discover ideas to differentiate your teaching and tasks to meet the needs of different levels of learners
- gather problems and discussion starters related to the math content
- identify critical vocabulary to help students better understand the content and strengthen their math communication skills
- reflect on your current teaching practices to expand your repertoire of teaching skills.

This book provides you with insights into the math expectations for kindergarten students; provides a wealth of standards-based activities; highlights research-informed instructional strategies; shares teacher notes that spur insights into classroom decision making; and provides a road map for teaching kindergarten students how to learn math, feel confident in their skills, and be excited to continue their math explorations in first grade. Read it, highlight it, and keep it close by as you and your students explore math together.

MODULE
1

Counting and Cardinality *Numbers 1–5*

About the Math K.CC.3; K.CC.4; K.CC.5

Although counting objects may seem like a simple math skill, there are many big ideas about numbers that help students understand the counting process. Counting is about more than simply saying the number names in the correct order.

- Do students understand that as they count objects, each number they say is paired with one object they touch (one-to-one correspondence)?
- Do students understand that if they touch and count (one-to-one), the last number they say is how many items are in that group (cardinality)?
- Do they understand that 4 is 4 no matter how the objects are arranged (conservation)?

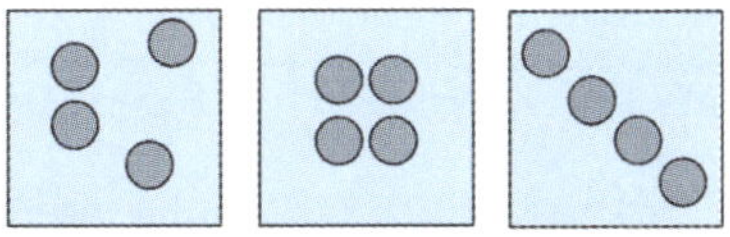

- Do they understand that when they count, each number represents 1 more than the number before it?

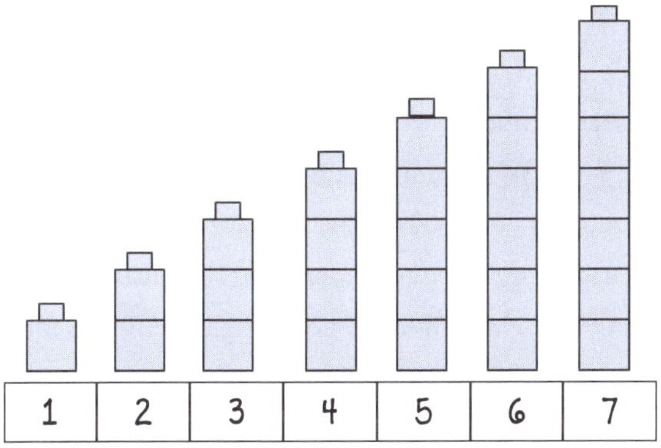

- Are they able to recognize the written numbers and write numbers that represent sets of objects?

Students develop these understandings through many hands-on experiences throughout the kindergarten year. Students visualize numbers with counters, five frames, and connecting cubes and observe and discuss number concepts.

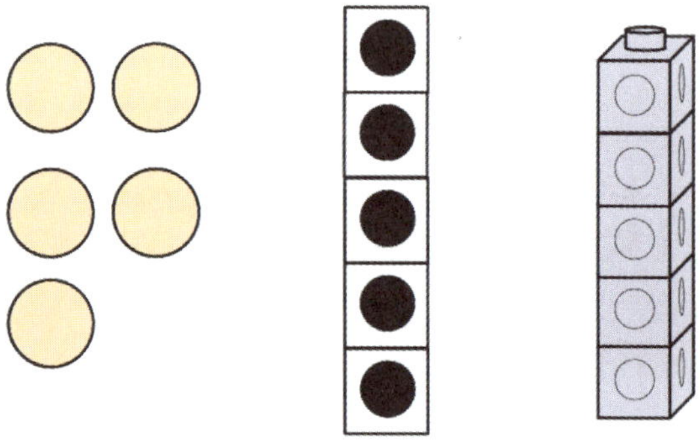

Although students may enter kindergarten being able to count to 5 and identify the numbers 1–5, they may not fully understand what these numbers represent. Although the counting order is discussed in Module 4, this module explores the development of understanding about the numbers 1–5 and the ability to count objects to 5.

The key ideas focused on in this module include:

- **counting up to 5 objects by touching or sliding**
- **saying 1 number name as each object is touched**
- **naming the total objects in a group of up to 5 objects**
- **counting 5 objects in different configurations**
- **recognizing the numerals 1–5**
- **selecting and writing the numeral that matches a group of objects**
- **recognizing that in the counting sequence each number said is 1 more than the last number said.**

Learning Goals

I can count up to 5 objects no matter how they are arranged.

I can match the numbers 1–5 with a group of 1–5 objects.

I know that when I count, every number I say is 1 more than the number before it.

I can write the numbers 1–5.

Counting 20 or fewer objects is a goal for kindergarten students. Because this task contains multiple skills (cardinality, conservation, one-to-one correspondence, reading and writing the numbers), we have chosen to break it into three modules (counting 1–5, 0–10, 0–20) to share instructional ideas and activities in a progression. Although you don't actually count 0 objects, we have included it with the counting 1–10 module because students need to understand the meaning of 0 and the symbolic representation for 0. Our goal, in all of the counting modules, is for our kindergarten students to be well grounded in the ideas of counting and cardinality.

Exploring the Progression

PREVIOUS		NOW		NEXT
Preschool		**Grade K**		**Grade 1**
Rote counting to 5; possibly counting five objects and identifying numerals to 5	»	Counting from 1–5 (including concepts of one-to-one correspondence, cardinality, conservation); writing numerals 1–5	»	Counting objects using tens and ones; reading and writing numerals to 120; representing 1–120 objects with the correct number

Ideas for Instruction and Assessment

COUNTING 1–5 OBJECTS

Students explore counting 1–5 objects by touching or sliding the objects, saying one number name as each object is touched and naming the total number of objects in the group. In addition, they begin to consider the relationship between numbers in the counting sequence (each number is 1 more than the number said just before it).

Introducing Numbers 1–5

Students explore the number 1 with actions, objects, drawings, and discussions.

Draw a dot card on the board with 1 dot inside.

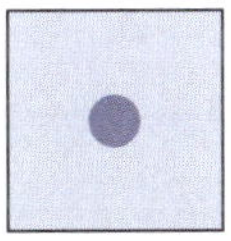

Turn and tell your partner how many dots you see. (1.)

Write the number 1 on the board.

What number is this? (1.)

How do you know it is 1?

Touch the dot and count "1."

Put a basket of colored square tiles at each table.

Show me 1 blue square.

Show me 1 yellow square.

Show me 1 red square.

Show me 1 green square.

Have students stand up.

Hop 1 time.

Turn 1 time.

Lessons in This Module

Counting 1–5 Objects
- Introducing Numbers 1–5
- Counting up to 5 Objects
- Exploring Five Frames
- 1–5 Caterpillars
- Five Frame Stickers

Counting Objects in Different Configurations
- People Count
- Exploring Different Arrangements of Counters
- What's My Number?
- Counting Different Arrangements of Dot Cards

Recognizing and Writing Numbers to 5
- *Rooster's Off to See the World*
- Flip and Write

Although many students have had previous exposure to the numbers 1–5, introducing each number to students allows them to talk about what each number means and to act out/draw their understanding of the quantity.

Repeat this lesson for the numbers 2–5 using similar activities.

Raise your hands 1 time.

Tap your head 1 time.

Give students paper and have them draw the dot card with 1 dot.

Have them draw 1 balloon (or 1 of any object) on their paper.

Have them label their drawing with the number 1.

Have them share their drawing with the class.

Thinking Through a Lesson

Counting up to 5 Objects

Students count manipulatives to 5.

Materials: bag of manipulatives with either 1, 2, 3, 4, or 5 items for each student

> Listen for accurate rote counting to 5.

> Watch as students count the cubes. Do they recognize the number of cubes without counting? Do they have one-to-one correspondence? Do they know that the last number they say is the total? Do they miscount due to a lack of one-to-one? Do they make an error in the counting sequence as they count (1, 2, 4, 5)?

Display the numbers 1–5 on the board or on a chart.

Have students count together from 1–5 while you touch each number on the board.

Show the students a bag of manipulatives.

How can we figure out how many bears/insects/cars are in each bag? (Count them.)

Give each student a bag of manipulatives.

Count to tell us how many are in your bag.

> If a student skips a number while counting the cubes, have them stop and begin again, listening to see if they make the same mistake in the counting sequence. Note which students need additional support with rote counting.

Listen to students as they count.

SMP3

Ask questions to highlight key ideas.

I noticed you were touching or pointing to bears while you counted. Why did you do that? (To know which ones I counted.)

What did you say every time you touched a bear? (A number.)

How did you know when to stop counting? (When there weren't any more bears.)

Have students share the number of objects in their bags.

> For students who need help verbalizing, try sentence starters to get them sharing about the number of objects in their bag. "I have (bears) in my bag. I counted the bears. I have ___ bears."

Have each student show how he or she counted the number in the bag by counting for you.

How many were in your bag?

How do you know? (It was the last number I said.)

Did anyone else have the same number of objects in your bag?

Switch the bags and repeat the process several times, having students count the objects and tell the number of objects in their bags.

Summarize the key ideas about counting.

Some of you lined up the bears before you counted.

Some of you touched each bear as you said each number.

You knew how many bears because it was the last number you said when you were counting.

SMP4

Notice the arrangement of cubes as students counted. Did the student line them up in a row? Did the student count them in a scattered configuration?

Differentiation

Listen as students practice counting to determine who might be having difficulty. As some continue to work with partners, spend time with the students who are having trouble with one-to-one correspondence, helping them touch each item as they count. For those who are having trouble with rote counting, have them say the numbers with a partner as they count.

FORMATIVE ASSESSMENT

Give a student 3 cubes and ask him to count them out loud.

Look for the student to demonstrate one-to-one correspondence.

As soon as the student counts the third cube, ask:

How many cubes are there?

Repeat with 5 cubes and then 4 cubes.

Record observational notes of student counting and responses.

If students are not able to count the cubes, record their errors (e.g., "Said 1, 2, 4, 5" or "Missed a cube").

If a student skips a number when counting, remove the cubes and ask her to rote count to 5. If the student is not able to count to 5, provide additional counting opportunities. Students can practice counting with a friend in different voices (e.g., whispers or silly voices) and count throughout the day during transitions.

When counting, if a student does not exhibit one-to-one correspondence, show her how to touch and slightly slide each cube as it is counted. This slide count helps students keep track of which cubes they have already counted.

Common Error

If a student needs to recount, he does not have cardinality.

An observational notes recording sheet is available in the online resources.

Exploring Five Frames

Students count objects to a quantity of 5 using a five frame.

Materials: five frame mats for students, a five frame mat for the teacher, 1–5 number cards, and bags of 1–5 objects to count

Show the group a blank five frame mat.

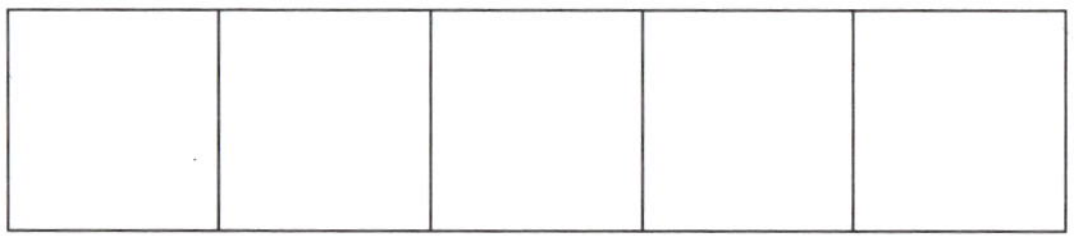

What do you notice about this frame?

How many boxes do you see?

This is called a five frame.

We are going to show our collections on five frames.

How could this help us count our collections?

During this introductory lesson, the five frame is shown horizontally and the frame is filled from left to right to model organization for the students; however, in future lessons it is important to show the five frame in different ways (e.g., the frame can be positioned vertically or students can fill it from right to left or show different arrangements and the number represented is the same).

Use the large five frame to model for the group how to count a collection of 3 bears by filling the frame from left to right, placing a bear in each consecutive spot.

Then, count the number of bears by touching each bear as you count.

Give each student a five frame and collection bag.

Show how you can count your collection on the five frame.

Allow students time to work.

How many objects are in your bag?

How do you know there are that many objects?

Students may need to talk about how they identified the total number of items in the bag (the last number they said when they counted).

Place number cards in front of the students.

Choose a number card to show the number of objects you counted.

How did the five frame help you count the objects?

Switch the bags and repeat the process several times, having students count the objects using the five frame and select the number card to match their set.

As an extension, give the students a number and have the students show that number on the five frame.

This student shows numbers on a five frame. He places counters on the frames to represent each number.

1–5 Caterpillars SHOW IT!

Students use yellow hexagon pattern block pieces to build caterpillars that show that each number is 1 more than the previous number.

This lesson can be modeled using an interactive whiteboard and clip art yellow hexagons.

Materials: 15 yellow hexagons and a set of number cards (1–5) for each student

Give each student a set of number cards and a set of hexagons.

Show me the number 1.

Observe as students hold up the number 1.

We are going to use our yellow hexagons to build a caterpillar to match each number.

Show a yellow caterpillar that is 1 hexagon long.

Observe as students place 1 hexagon.

How many hexagons did you use for your caterpillar? (1.)

Model on the board a caterpillar that is 1 hexagon long. Write *1* next to it.

Place your number 1 with the number 1 caterpillar.

Build a caterpillar that shows 2 right under the number 1 caterpillar.

Model on the board a caterpillar that has been made using 2 hexagons. Write *2* next to it.

The hexagons should be lined up to show that each number gets bigger by 1.

Have students continue building caterpillars for 3, 4, and 5.

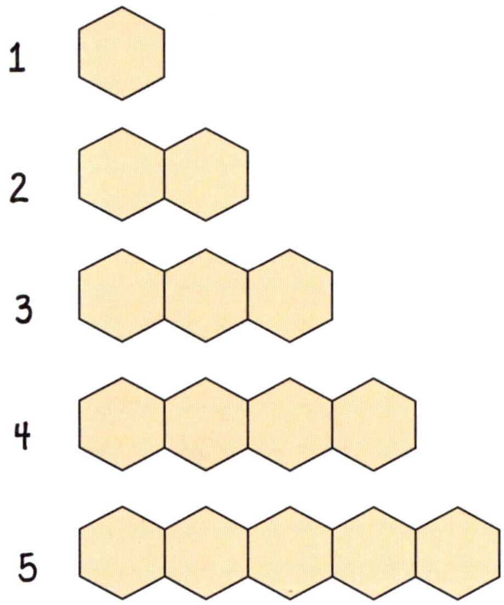

For more ideas on using models to build math understanding, see Math in Practice: A Guide for Teachers, Chapter 3.

What do you notice about the caterpillars?

Which is the smallest? Then which is a little longer?

Continue asking which one is a little longer until there is just one more, then ask:

Which is the longest?

Every time we added another block, what happened to our caterpillar?

What did you notice about the caterpillars that were 1, 2, 3, 4, and 5 blocks? How much bigger did they get every time we counted another number?

Five Frame Stickers

Students could also use bingo markers or crayons to show numbers in the five frames.

Students show the numbers 1–5 on five frames using stickers.

Materials: 5 five frame mats and small stickers that will fit in the five frames for each student

Give each student one blank five frame.

How can we use stickers to show 1 on this five frame?

Where would you put the one sticker?

The five frame should be horizontal, placing the sticker in the first box on the left. Continue to show other numbers by adding one to each box.

Model for the students by placing a sticker in the first box of the frame.

Allow students to place their sticker.

What number does this frame show? (1.)

Give the students another blank five frame.

What number comes after 1 when we count?

Where would you put the two stickers?

Allow students to place their stickers.

Give the students another blank five frame.

What number comes after 2 when we count?

Where would you put the three stickers?

Continue making five frames to show 4 and 5. When the frames are complete, ask students to put them in counting order.

Observe as students place the frames in order from 1 to 5.

Tell your partner something you notice about the number of stickers on the frames.

Hold up your frame to answer each of these questions:

What number is 1 more than 1?

What number is 1 more than 2?

What number is 1 more than 3?

What number is 1 more than 4?

What do you notice about each number as we count?

Have students write their name on the back of each frame and use the set to practice identifying the numbers (e.g., matching number cards to the frames or showing a frame to a partner and having him say the number or write the number for each frame).

Additional Ideas for Support and Practice

The following ideas extend students' skills of counting sets of objects and provide meaningful practice.

Counting Page by Page

As numbers are introduced, students create pages to a counting book. The pages include:

- the number
- a five frame that is filled in to show the number
- a square that becomes a dot card to show the number
- a drawing to show that quantity of anything the students would like to illustrate.

The pages are saved and stapled together to form a book later in the year.

Number Cups

Students use a set of cups and practice counting by putting the number of counters in each labeled cup.

Label plastic cups with numbers 1 through 5.

Students order the cups from 1–5, then count to put the appropriate number of counters in each cup.

Differentiation

If students are having difficulty with the number sequence 1–5, give them number strips or number lines to remind them of the sequence. Students can use these tools while doing centers or practice tasks, too.

Counting Collections Center

Students practice counting collections of various items. At this level, students need lots of practice counting objects.

Ideas for collections include pennies, beads, erasers, crayons, paper clips, or pom-poms.

Place collections of 1–5 objects in bags, cups, or trays in the center area for students to count.

Simon Says

Students show their counting skills and understanding of numbers through movement and a modified game of Simon Says.

Play a counting version of Simon Says.

Simon says clap and count to 3.
Simon says jump and count to 4.
Simon says march and count to 5.
Simon says twirl and count to 2.

Five Frame Apples

Students roll a 1–6 number cube to cover a five frame on an apple.

Materials: number cube (1–5, cover the 6 with a sticker), 5 white connecting cubes per student, 5 red connecting cubes per student, and a five frame apple per student

Player A rolls the number cube and places that many white cubes on the apple (e.g., rolls a 3 and puts 3 white cubes on the apple).

Then Player B rolls the number cube and places that many white cubes on the apple (e.g., 4).

Partners take turns rolling the number cube. They must roll an exact amount to fill the five frame.

If players roll the side of the die with a sticker, they roll again.

The player who fills the five frame exactly gets to take 1 red "apple" cube.

The red "apple" cubes can be placed beside the apple.

The player then removes all 5 white cubes from her apple and starts again.

Play continues until one player has 5 red "apple" cubes. That player is the winner.

Number Bracelets SHOW IT!

Students count using number bracelets made from pipe cleaners and beads. (It is helpful if the beads are a different color from the pipe cleaner.)

Form a circle with the pipe cleaner, put on 1 bead, and twist the ends together. Continue with numbers 2 through 5 using the same color of beads on each bracelet.

Students count the number of beads on a bracelet and match it to a number card as independent practice.

Sets of bracelets for numbers 1–5 can be stored on ring hooks for each student.

Differentiation

If students are having difficulty counting, pair them with students who know the counting sequence so they can hear them as they count.

These number bracelets show numbers 1–3. They can be kept in sets using a chart ring.

Material Hunt

In a small group, students select a number card and an object card and find that number of items in the classroom.

Materials: picture cards of plentiful objects in the classroom (glue sticks, manipulatives, crayons, etc.) and number cards (1–5) for each group

In a small group, one student picks a picture card and a number card (e.g., the number 3 is picked along with a picture of bear manipulatives).

Each student in the group finds 3 bear manipulatives.

Have students self-check for accuracy by counting their objects.

Then the next student selects a picture and a number card to show the item for everyone to gather.

Classroom Number Towers

As numbers are introduced throughout the year, the teacher adds each number to the classroom wall to create a number tower. The numbers are posted left to right and graphics (e.g., a bear, butterfly, heart) are placed above each number to show the quantity. The teacher creates the tower with students' input, asking them to tell how many graphics to add and counting as she adds them.

As more numbers are introduced, students are able to see the written numeral connected to the visual, and the relationship of each number to the next consecutive counting number becomes apparent. Students and teacher can refer to the number tower throughout the year when counting, identifying numbers, or even comparing numbers.

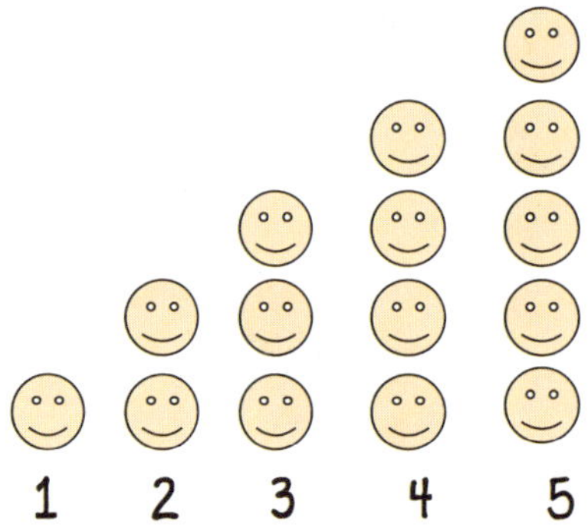

Pattern Block Numbers Center

Students use sets of the same-pattern blocks (e.g., red trapezoids) to show the numbers 1–5.

Materials: sets of same-pattern blocks and number cards

> Students use red trapezoids to build electric eels to show numbers 1–5 (e.g., a student builds an eel with 1 red trapezoid, then an eel with 2 trapezoids, and continues with 3, 4, and 5).

Students place them in order 1–5 to see how the length increases.

Then students label the eels with number cards.

Students might also use pattern blocks to build green triangle snakes and orange square worms.

COUNTING OBJECTS IN DIFFERENT CONFIGURATIONS

Students apply their counting skills to count objects in lines, circles, arrays, and scattered configurations.

People Count

Students count groups of friends in different arrangements.

Have 4 students stand. Put them in a row.

How many people are in the row?

Count them out loud.

Arrange them in a circle.

How many people are in the circle?

Count them out loud.

Have the 4 students move to the 4 corners of the room.

How many people are in the corners?

Count them out loud.

How do we know there are still 4 people?

Have those students sit and invite 3 students to stand.

Repeat with different numbers of students, moving them into different configurations.

Exploring Different Arrangements of Counters SHOW IT!

Students practice counting manipulatives in various configurations.

In a small group, set out a few scattered arrangements of 4–5 counters.

Have each student work with a partner.

Ask one partner to count them to the other partner.

Notice how the students count the sets.

Notice positive strategies like these:

- rearranging the counters in a line and then counting
- counting in a systematic way, left to right or top to bottom

Subitizing is instantly recognizing a quantity. For example, the arrangement of the number 4 on a die.

Differentiation

Students may need each of the strategies modeled through small-group instruction.

- touching or sliding each counter as they count
- starting to automatically recognize the count (subitize).

Notice potential errors like these:

- counting in a haphazard way (not systematic)
- counting an object twice
- missing an object altogether.

Pose questions based on what you see:

Sam, why did you drag the counters?

Claire, why did you put them in a row?

Was it hard to count them when they were scattered around? What could you do to make it easier?

How could you be sure not to miss any?

What's My Number?

This lesson is best done in small groups to allow you to observe students as they count and hear their explanations.

Students count manipulatives in various configurations.

Materials: sets of manipulatives with 3–5 items in each bag (i.e., each student will need a bag with 5 items, a bag with 4 items, and a bag with 3 items)

Give each student a bag of 5 bears (bugs, cars, or other manipulatives).

Have students count their bears.

Have them place their bears in a row and count them again.

How many?

Does that make sense? Why?

Common Misconception

Sometimes students think, because the arrangement of the cubes changed, that the total number of cubes changed.

Have them place their bears in a diagonal and count them again.

How many?

Have them scatter the bears and count again.

How many? What do you notice?

Switch bags and have students line up the manipulatives in different ways, counting them each time.

One student counts a line of frogs and the other counts a line of bears. The girls touch each item as they count.

Counting Different Arrangements of Dot Cards

Students count the dots on cards and match them to the number shown.

Materials: a set of dot cards 1–5 and a set of number cards 1–5 for each student

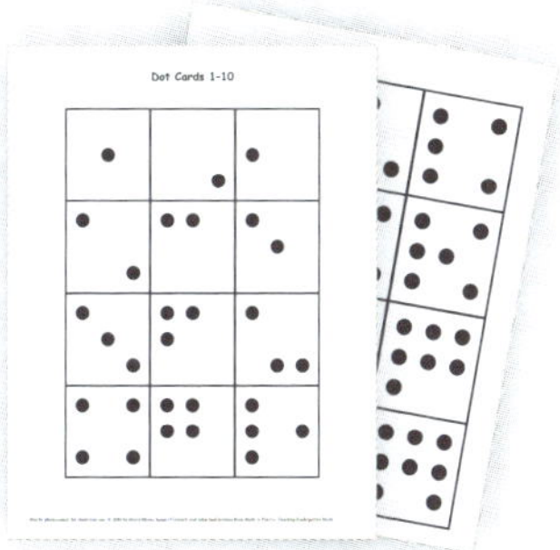

Have students work with partners.

Give partners the number cards for 1–5 plus a set of dot cards.

How can we figure out how many dots are on each card?

Dot cards for 0–10 are available in the online resources. Select cards with 1–5 dots for this activity.

Have students order their number cards in a row, 1–5.

Have each student choose a dot card from their pile of dot cards.

How many dots are on your card?

Observe as students count.

Turn and talk to your partner about how many dots are on your card.
Put your dot card by the number card it matches.
Select another dot card. How many dots are on it?
Match it to the number card.

Students should count with one-to-one correspondence. As they count, notice whether they say one number for each dot.

Continue placing dot cards with the number that matches.

When partners have two cards that show the same number, ask:

How are these cards the same?
How are they different?

After partners have counted the dots and sorted the cards by number, say:

Hold up a dot card that shows 1.
Look around the group at the cards, what do you notice?
Hold up a dot card that shows 2.
Are all of the 2 cards the same?

There will be three dot cards that match each number.

Show examples of 2.

Continue with other numbers to share different ways each number can be shown with dots.

FORMATIVE ASSESSMENT

Arrange 4 cubes in a vertical line.

Count the cubes.

A student may touch each cube as she counts or slide count by slightly moving each cube.

As soon as the student counts the fourth cube, ask,

How many cubes are there?

Then place the 4 cubes in a different arrangement (e.g., scattered configuration or a horizontal line).

How many cubes are there?

The student should be able to say that there are 4 cubes. If the student needs to recount, he may not understand that the arrangement of the cubes does not change the number of cubes.

Numbers 1–5
Student Observation Notes

Record observational notes of student counting and response.

Continue with 5 and then 3 cubes.

Some students may be able to count cubes in a line but have difficulty counting a scattered arrangement. These students should be taught to touch and slightly slide each cube as it is counted. The slide count helps students keep track of which cubes they have already counted.

Additional Ideas for Support and Practice

The following ideas extend students' understanding of counting sets of objects in different configurations and provide meaningful practice.

Talk About It/Write About It

Have students show 4 in several ways using manipulatives.

How do you know they are all 4?

This can be adapted for any number.

Have students draw four ways to show the quantity 4 in their journal and label the drawings with the number 4.

Number Pairs

Students work with a partner to show numbers 1–5 on a five frame and another way using counters.

Materials: a five frame, 10 counters (5 for each partner), and a set of 1–5 cards for each pair

Pick a number card 1–5.

Player A shows the number on a five frame.

Player B shows the number in a different arrangement on the table.

Students count to check their partner.

Then switch roles and pick another number.

Arrange the Buttons

Using large buttons, students make different arrangements of numbers 2–5 and trace them.

Choose a number 2–5 to focus on for the day and have each student in a small group select that number of buttons.

On a piece of paper or in math journals, create four quadrants for students to work in (see below).

The students place all the buttons in one quadrant and trace them.

Then they will move the buttons to the next quadrant in a different arrangement and trace again.

Students do this four times, each time making sure to arrange the buttons differently.

Students label the number represented in each quadrant and continue on a new page with a different number of buttons.

Differentiation

If students struggle tracing the buttons, they can draw buttons or use stickers to make different representations in each quadrant.

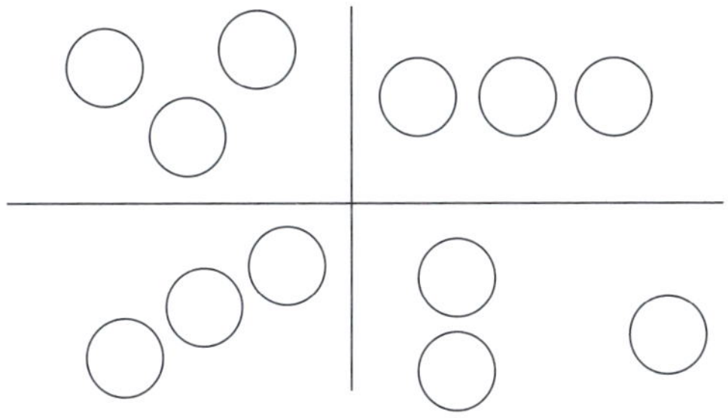

Silly Sticks

Students count and arrange craft sticks.

Select a number card (1–5) and give each student that number of craft sticks or plastic stirrers.

Seat students at a table or on the carpet so they can arrange the sticks in any way they wish and the arrangements are visible.

Look around the carpet or table and talk about how the different arrangements look.

Do any look alike? Do any form a shape?

Have students touch and count their sticks.

How many sticks are in each silly shape?
Can you show me 3 in a different way?

Acknowledge that the sticks may look different but represent the same quantity.

Designate a different number of sticks and have students make silly shapes with that number of sticks.

RECOGNIZING AND WRITING NUMBERS TO 5

Students continue to count objects and connect their counts to numbers, but these lesson ideas emphasize the written component in which students either select or write the number to match the total objects in the group.

Rooster's Off to See the World

Students use copies of the animal templates from the story to count each type of animal and write the number of each animal on a dry erase board.

Materials: copies of the animal templates copied from inside the book, individual dry erase boards, and markers

Before Reading:

Show students the cover of the book.

We are going to read a story about a rooster who takes a trip with friends. Listen to see what friends join him on his trip.

During Reading:

Read *Rooster's Off to See the World* by Eric Carle.

As you read, count the number of animals on each page.

As animals join or leave Rooster on the trip, ask students if there are more animals or fewer animals.

Students explore these terms in greater detail later, but can be exposed to the concept of more or fewer as they explore numbers.

After Reading:

Give students copies of the animal templates, dry erase boards, and markers.

Have students sort the templates by animal.

How many roosters were in the story? Count the number of roosters.

Show how you can write the number 1 on your board.

Observe as students write.

Hold up your boards to show the class your number 1.

Where do we see the number 1 in our classroom?

Have students erase their boards.

Count the cats in your set. How many cats were in the story?

Show how you can write the number 2 on your board.

Observe as students write.

Hold up your boards to show the class your number 2.

Where do we see the number 2 in our classroom?

Continue the lesson by counting 3 frogs, 4 turtles, and 5 fish, having students write each number to match the number of animals.

Identify examples of the numbers found in the classroom.

Differentiation

Some students may benefit from having a number strip to see the number closely before writing it.

Take note of students who are writing numbers backward. Have them check the formation of their number with the 2 on the number strip. If needed, pause to revisit the correct formation of the number.

Flip and Write

Students flip a dot card, count the number of dots, write the number to match the card, and check their partner's work.

Materials: dot cards for numbers 1–5 for each pair of students, individual dry erase boards with marker (or paper and pencils) for each student

Give each pair of students a set of dot cards and each student a dry erase board and a marker.

Make a stack of dot cards and turn it upside down so we can't see the dots.

Flip over a card from your stack. Don't let your partner see your card.

Count the number of dots on your card and then write the number on your board.

Watch as the students count the number of dots.

Observe that the number written matches the number of dots.

If you have trouble remembering how to write a number, where can you look for help?

Let's check to see if you counted correctly and wrote the number to match.

Switch dot cards and boards with your partner.

Check to be sure that your partner counted correctly and wrote the

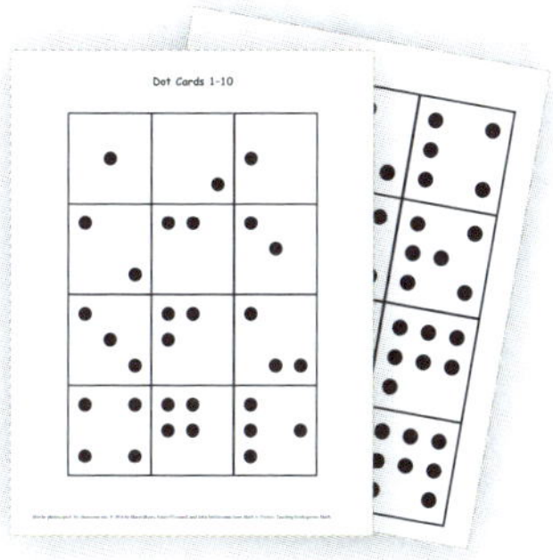

Dot cards for 0–10 are available in the online resources. Select cards with 1–5 dots for this activity.

right number.

What number did your partner write?

Does it match the number on the card?

How did you count the number of dots?

Continue having students flip a card, count the number of dots, write the numeral to match, and then check with partners.

FORMATIVE ASSESSMENT

Ask students to draw 4 balloons and label their picture with the number 4.

This student drew 4 balloons and was able to write the number 4.

At this level, reversals are expected. Continue to remind students of the proper formation and give them opportunities to practice.

If they are having difficulty with a specific number, tape the number to their desk or use a yellow highlighter to write the number for them and have them trace over the highlighter to practice forming the number.

Additional Ideas for Support and Practice

The following ideas extend students' understanding of the written numerals 1–5 and provide meaningful practice.

Stamp It

Students show quantities of 1–5 using stamps and write the number to match.

Place number cards 1–5, paper, pencils, a variety of stamps (or stickers), and ink pads at a center.

Students pick a number card 1–5.

They use stamps and ink pads to show that number (e.g., showing 4 by stamping 4 hearts).

Students then label their picture with the number.

Each student's pages can be put together to make her own number book.

Building Numbers on a Five Frame

Students show the numbers 1–5 on five frames and write the number to match.

Put copies of a blank five frame in a sheet protector.

Students use the sheet protector as a dry erase board by writing on it with a dry erase marker.

Students draw 1 dot on the five frame and write the number 1 to label the number shown.

Then, they erase their board and draw 2 dots and write the number 2.

Continue with numbers 3–5.

Clothesline Ordering

Students match dots shown on clothespins to number tents on a clothesline.

Materials: string, 5 clothespins with 1–5 dots on each, number tents 1–5 (5 × 10-inch strips of cardstock paper folded in half to form 5 × 5-inch tents with one flap labeled 1–5)

Tie a string low enough for students to reach to make a clothesline.

Have students place the number tents on the string in order from 1–5.

Students then choose a clothespin, count the number of dots, and place it on top of the matching number tent.

Number Dot Cards

To support students who struggle with the recognition of numbers and/or with counting numbers, consider using cards with dots to help them connect the numeral and the quantity.

A student makes sets of cubes to match the number of dots on numeral cards. Seeing the number with dots helps the student learn the numbers.

Paint Numbers

Allow students to practice forming numbers by filling a gallon ziplock bag with some acrylic paint and having students trace the numbers over the bag.

Students might roll a dot cube and then trace that number on the paint bag or pick a dot card and then trace that number, giving them practice with counting the dots and forming the numbers.

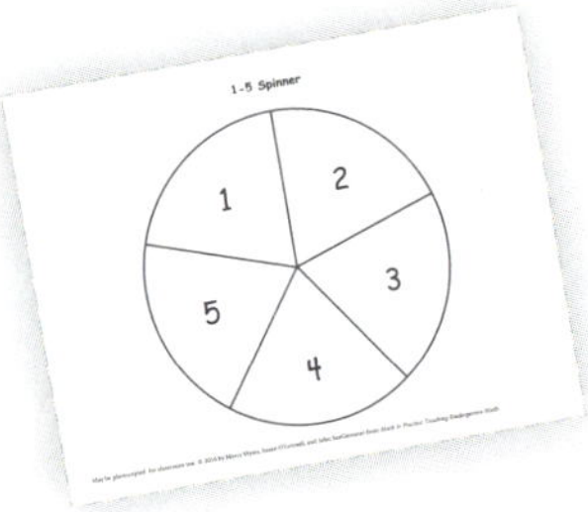

Climb to the Top

Students spin a 1–5 spinner and then write the number to see which number wins.

Materials: 1–5 spinner and Climb to the Top 1–5 recording sheet

Before beginning, students predict which number will make it to the top first.

Students then spin a number and write the number spun in the box on the recording sheet beginning at the bottom of the column.

Students continue spinning and writing the number until all of the spaces for a number are filled; that number reached the top first. They color the happy face to show it was the winner!

This student uses a number cube to practice writing numbers 1–5. This activity can be modified for any numbers by using different spinners or dice and can be done independently or as a partner game.

I Say, You Write

Students practice writing numbers with a partner.

Player A says a number, 1 to 5.

Player B writes the number on a dry erase board.

Player A checks the number written, then the partners switch jobs.

The connections between the counting standards is evident in this activity as students write numbers, draw to show quantities of numbers, and talk about their insights about each counting number being 1 more than the number said before it.

Vocabulary

arrangement
collection
counting numbers
dot card
five frame
number card
number cubes
set

ONLINE RESOURCES

General Resources

Additional Problems

Dot Cards 0–10

Five Frame Mats

Number Cards 1–5

Numbers 1–5 Student Observation Notes

Resources for Specific Activities

1–5 Spinner

Climb to the Top 1–5 Recording Sheet

Five Frame Apple

These resources are available at http://hein.pub/MathinPractice, keycode MIPGK.

Folded Book

Students make accordion books in which they write and show each number and then talk about each number being 1 more than the number just before it in the counting order.

Fold construction paper accordion-style to make a 5-section book.

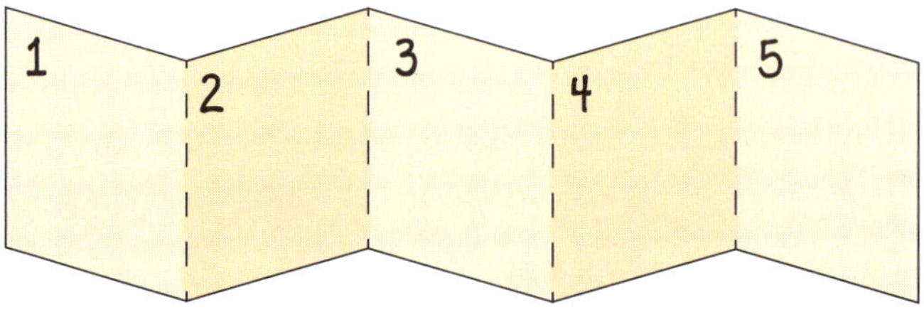

Students write the number 1 and put 1 sticker (or draw 1 animal) in the first space.

Continue with numbers 2–5.

At group time, students bring their books to share and talk about what they notice about the numbers (each number showing 1 more animal or sticker than the number just before it).

MODULE

2

Counting and Cardinality *Numbers 0–10*

About the Math K.CC.3; K.CC.4; K.CC.5

In this module, students continue their explorations with counting up to, and including, 10 objects.

The key ideas focused on in this module include:

- using counting and sliding strategies to count up to 10 objects in different configurations
- naming the number of objects in a group of 0–10
- recognizing and writing the written numerals 0–10
- recognizing that in the counting sequence each number said is 1 more than the last number said.

Again our focus is on the big ideas of counting and cardinality.

- Do students understand that, as they count objects, each number is paired with one object (one-to-one correspondence)?
- Do students understand that if they touch and count (one-to-one), the last number they say is how many items are in that group (cardinality)?
- Do they understand that 8 is 8 no matter how the objects are arranged (conservation)?
- Do they understand that when they count, each number represents 1 more than the number before it?

Lessons in This Module

Counting Objects to 10

Introducing Numbers 6–10
Building a Ten Frame
Counting Cubes
Show Me
Counting Stickers
Zero Is the Leaves on the Tree
Number Towers
Ten Black Dots

Recognizing and Writing Numbers to 10

Counting Bags of Cubes
Writing Numbers 0–10

Learning Goals

I can count 10 or fewer objects no matter how they are arranged.

I know that when I count, every number I say is 1 more than the number before it.

I can match the numbers 0–10 with a group of 0–10 objects.

I can write the numbers 0–10.

- Are they able to recognize the written numbers and write numbers to represent sets of objects?
- Are students able to immediately know how many objects are in a group without counting (subitizing)?

The hands-on activities in this module build on those in the previous module. Now students count 10 or fewer objects and recognize and write numbers 0–10. Students visualize these quantities with counters, ten frames, number lines, and connecting cubes and continue to observe and discuss number concepts. (See Figure 2.1.)

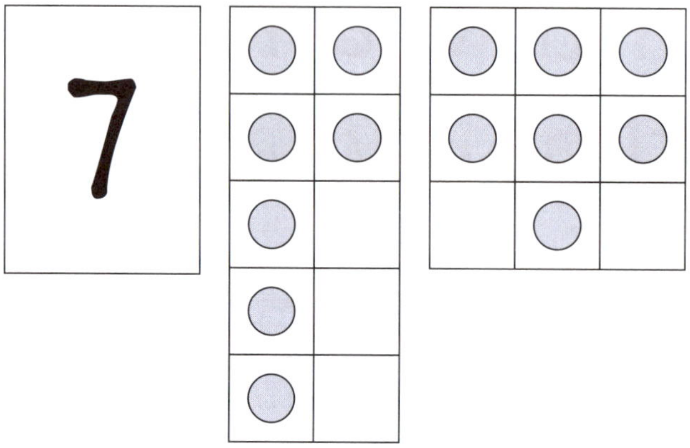

Figure 2.1 Students explore numbers using various models to build a deeper understanding of number concepts.

Exploring the Progression

PREVIOUS		NOW		NEXT
Preschool		**Grade K**		**Grade 1**
Rote counting to 10; possibly counting objects and identifying numerals to 10	»	Counting from 0–10 (including concepts of one-to-one correspondence, cardinality, conservation); writing numerals 0–10	»	Counting objects using tens and ones; reading and writing numerals to 120; representing 1–120 objects with the correct number

Ideas for Instruction and Assessment

COUNTING OBJECTS TO 10

Students practice counting objects using touching and sliding strategies to show their understanding of one-to-one correspondence and explore counting objects in different configurations.

Introducing Numbers 6–10

Students explore numbers 6–10 with actions, objects, drawings, and discussions.

> Introduce students to numbers 6–10 by counting, acting out, and drawing the quantities. This lesson shows examples for 6, but can be modified for any quantity.

Give each pair of students a bag with 6 cubes in it.

Work with your partner to decide how many cubes you have.

Write the number 6 on the board.

Put a basket of pattern blocks at each table.

Show me 6 yellow hexagons.

Show me 6 green triangles.

Show me 6 orange squares.

Show me 6 red trapezoids.

Have students stand up.

Clap 6 times.

Jump 6 times.

Give your partner 6 high fives.

Give students a paper and have them draw 6 of something.

Have them label their drawing with the number 6.

Have them share their drawing with the class.

Building a Ten Frame

Students are introduced to ten frames and count objects on a ten frame.

Give each student a five frame and 5 counters.

Show me 5 on your frame.

How do you know it is 5?

Have students work with partners to slide each of their five frames together to make a ten frame. Model doing it on the board.

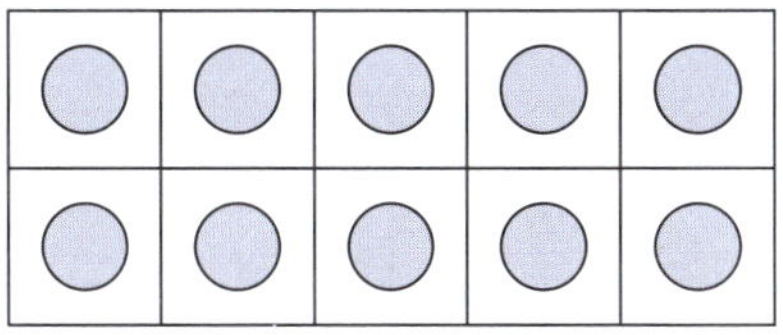

What number do your frames show now that you have put them together? Explain.

Give students a ten frame.

How many counters fit on this frame? Try it to see.

Does that surprise you? Does it make sense?

Some students may need to touch and count each box.

Have students rotate their ten frame to a vertical display.

How many dots do you see?

How do you know?

Repeating the same questions that ask students how they arrived at numbers is important as students learn to subitize numbers on ten frames.

Display a horizontal ten frame showing 5, with the dots filling the top row.

How many dots do you see?

How do you know? (It looks like a five frame.)

What do you notice about the top row?

What do you notice about the bottom row?

How could you check to be sure it is 5?

Students should understand that there are still 10 boxes on the frame. The position of the ten frame does not matter. This helps with conservation of number and builds an understanding that a ten frame always has 10 boxes.

Turn the frame to show 5 vertically and ask the same questions, being sure that students understand that it is still 5.

Display a ten frame showing 7 dots, filled left to right beginning with the top row.

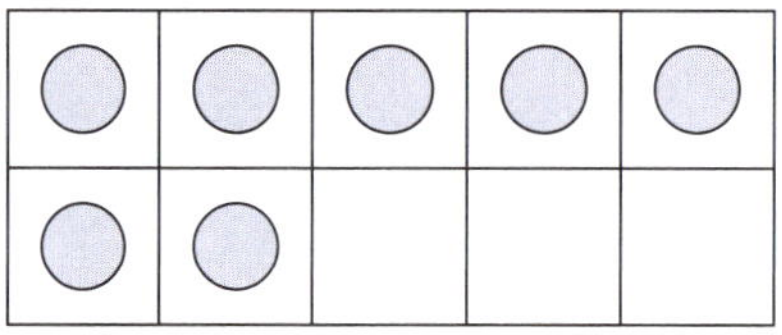

How many dots do you see?

How do you know?

Some may need to count each dot but others may see 5 on the top and be able to count on 6, 7. Other students may see the group of 2 on the top and 2 on the bottom to make 4 and then count on the remaining 3 on top as 5, 6, 7.

Students may notice that there are 3 empty boxes on the bottom.

Continue with different numbers over several days. Make connections to numbers and how the frame "fills up" as the numbers increase.

Show 7 vertically on the frame and have students discuss and justify that it is still 7 dots.

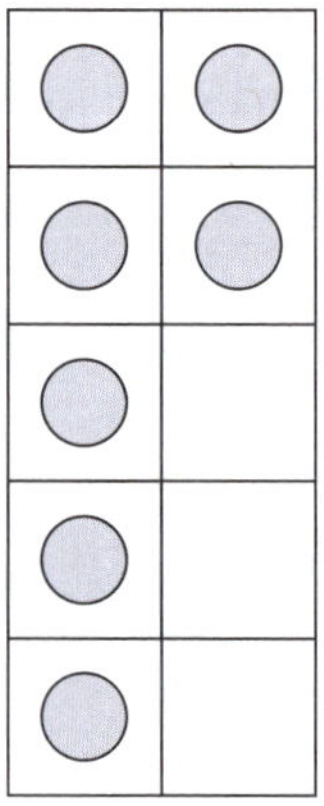

Counting Cubes

In this small-group activity, students count cubes in different configurations to discover that a number of objects stays the same no matter how it is positioned.

Arrange 7 cubes in a horizontal row on the table.

How many cubes are there?

Ask a student to count the cubes. Watch to see that the student touches, slides, or pulls each item as he counts and says each number.

Take the 7 cubes and put them in a vertical column on the table.

How many cubes are there?

Ask a different student to count the cubes.

Watch to see that the student touches, slides, or pulls each item as she counts and says each number.

Take the 7 cubes and put them in a circle on the table.

How many cubes are there?

Ask a different student to count the cubes, again watching for counting skills or the ability to already know that there will be 7.

Take the 7 cubes and scatter them on the table.

Turn and tell your partner: How many cubes are there?
How did you know?
But it looked different. Does that matter?

Have each student take 6 cubes from a basket of cubes.

Show me 6 cubes in any way you want.
Look at your partner's cubes. Do they look the same as yours?
Count your partner's cubes. Are there 6?

Watch for students who continue to need to count the cubes even though they are seeing that you have not added or taken any cubes from the group.

FORMATIVE ASSESSMENT

Arrange 6 cubes in a line for a student to count.

Count the cubes.

Observe as students count with one-to-one correspondence, saying each number in the counting sequence.

As soon as the student counts the sixth cube, ask,

How many cubes are there?

To demonstrate cardinality, the student should state that there are 6 cubes without recounting the group.

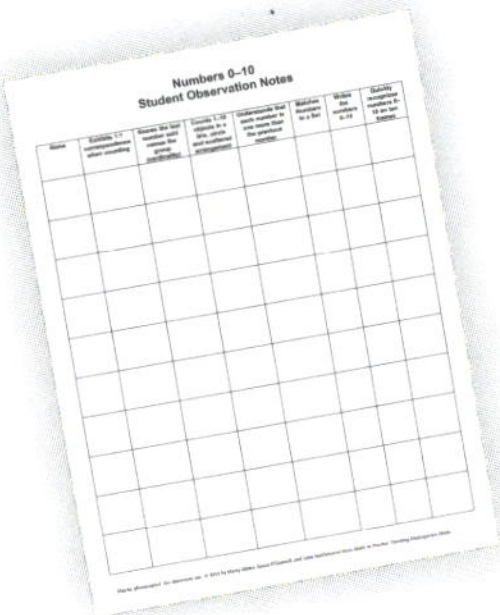
Numbers 0–10
Student Observation Notes

Then, move the cubes to a different configuration.

How many cubes are there?

The student should recognize there are still 6 cubes because none were added or taken away.

Record observational notes of student counting.

Continue with 8, 7, 9, and 10 cubes.

Students should be assessed counting numbers 6–10 in line, array, circle, and scatter arrangements.

Use the observational notes page.

Assessment is ongoing as students practice counting.

Differentiation

If students are not able to count groups of 6–10 items, reassess their counting of groups to 5. Then continue to practice with numbers to 10.

Observing students as they count is important to assess one-to-one correspondence and cardinality. The teacher watches as the student counts the group of cubes by touching each one. To assess cardinality, she asks, "How many cubes are there?" right after the student counts the last cube.

Show Me SHOW IT!

Students explore using the ten frame to show quantities to 10 by placing counters on their ten frame and explaining how they did it.

Give each student a ten frame.

Show me 4.

How do you know you have 4?

Show me 9.

How do you know you have 9?

Show me 6.

How do you know you have 6?

I see some different ways some of you are making 6.

Show a couple different ways to place 6 counters on the frame like the ones below.

Ask students to share their explanations with the class. Hearing explanations from other students can deepen students' understanding of numbers.

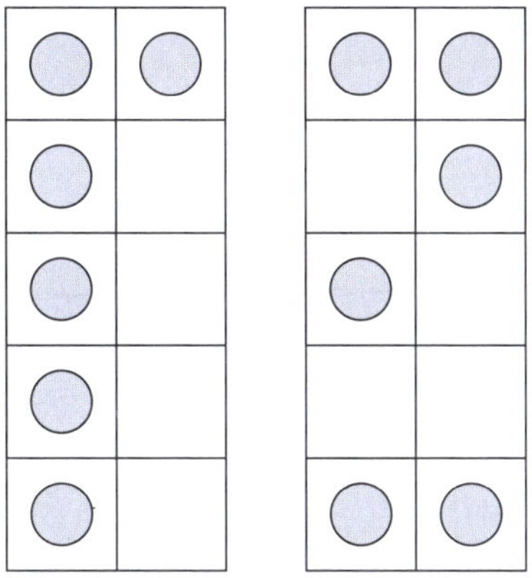

For each, ask students if it shows 6 and how they know it is 6.

Show another way to place the counters on the frame that would still be 6.

Watch as students model 6 on their frames.

Let's see how you showed 6 in different ways.

Have students share a few more examples.

This is a good opportunity to watch for one-to-one correspondence as students count to prove there are 6 on the frame.

How can you prove that you showed 6?

Do all of these show 6? Explain.

6 is still 6 no matter where the counters are placed.

How does it help you to fill all 5 in a row first?

We can show 6 in lots of ways and it is still 6, but for some ways we might have to count to know it is 6.

Continue to pose numbers for students to show on the frame, watching for their counting skills and their understanding that a number doesn't change because it is in a different configuration.

Counting Stickers

Students count various arrangements of stickers for numbers 6–10 and learn to keep track of the count by marking each one as they count.

Materials: sheets of paper with stickers in various line, circle, and scattered arrangements for the numbers 6–10, sheet protectors, dry erase markers

Give each student in a small group a sheet protector and a page of stickers in a row.

How can we figure out how many stickers there are on this page?
Count to find out how many stickers are on your page.

As you observe, take note of students who lose track of how many stickers they are counting and count too few or too many, students who count accurately by touch counting, and students who count accurately without touch counting.

Watch as students count.

Show us your page and tell us how many stickers you counted.
How did you count them, since the stickers can't be moved?
How can you know where you started?
How will you know what you have already touched?

Give each student a new page of stickers, this time in a circle or scattered configuration.

Count to find out how many stickers are on your page.

Common Error

Students may have difficulty keeping track of the count for scatter and circle arrangements. They may not remember where they started the count. Help students learn how to keep track of their starting point as they count.

Watch as students count.

After students finish counting, say:

Show us your page and tell us how many stickers you counted.
How did you count them, since the stickers can't be moved?
How did you know where you started? Did you get confused?
How did you know what you already touched?

Allow students to talk about how this might have been more difficult than counting the stickers in a row and how they got confused.

Differentiation

Some students may be ready to identify groups of stickers to combine to find the total number rather than counting each individual sticker. For example, they will see a group of 3 and a group of 2 to make a total of 5.

Give each student a dry erase marker.

How could you use this marker to help you as you count?
Try it. See if it makes it easier for you to count.

Have students recount the number of stickers by marking each one.

How many stickers did you count?
How do you know?
Was it easier when you crossed off as you counted? Why?

Have student erase the marks on their page.

Give them a new page to count the number of stickers.

Students continue to make marks to keep track of the count.

Thinking Through a Lesson

Zero Is the Leaves on the Tree

Students are introduced to the concept and symbol for 0 through a story and then brainstorm examples to show their understanding of 0.

Students at this level may not understand what 20 or 100 or 1,000 represents, but may recognize them as large quantities. With their current understanding of the numbers 1–10, we would expect them to choose numbers closer to 10 when describing how many they'd like to have.

Before Reading:

Ask students to turn and tell something they like a lot.

Have them share some of the things they like (e.g., toys, cookies, kites, balloons, bubbles, dogs).

How many would you like? Why?

They will likely want large quantities like 9 or 10 or may even say 20 or 100 or 1,000.

Would you like 0 of them? Explain.

SMP3

Some students may get that 0 means none of them, but others may not.

Although students have heard the term zero, they may not understand that it means having none of something.

Tell them you are going to read a story about 0.

During Reading:

Read *Zero Is the Leaves on the Tree* by Betsy Franco.

As you show each picture, ask students if they can count the number of balls or leaves or ducks.

After Reading:

What does it mean to have 0 of something? (You don't have any.)

Tell students they are going to draw a picture, like in the book, that shows 0 of something.

How did the illustrator of the book show 0 in drawings? (She drew no leaves on the tree. *Or,* She didn't put any kites in the sky. *Or,* There were no bikes on the rack.)

Show the pictures from the book again as students remember the examples of 0.

Turn and talk: What could you draw that shows 0?

Can you have any of them in your picture?

Differentiation

If students have a difficult time coming up with examples, ask them specific questions like: If you drew a picture of 0 scoops of ice cream on a cone, what would it look like? If you drew a picture of 0 apples on a tree, what would the tree look like?

Have students share some ideas.

Have each student draw a picture to show 0 and record 0 on the page. Observe as they draw. Have each student share his or her picture.

Tell us what you have 0 of in your picture.

Support any students who may have misunderstood the meaning of 0 with questions like, "Can you have any dogs on your picture if there are 0 dogs? Why?"

Refer back to the first page of the book. (The teacher in the story wrote *0, 1, 2, 3, 4, 5, 6* on the board.) Write the numbers on the board in the same way as in the book and ask students where 0 is in the numbers on the board.

SMP3

Students may not have seen 0 in the counting order because hundred charts, number strips, and other classroom tools tend to begin with 1.

Turn and talk: Why did the teacher put 0 there? (You don't have any if you have 0 and with all the other numbers you have some. *Or,* You have more if you have 1 or 2 or 3.)

Have students summarize their understanding by talking with partners:

What does it mean to have zero of something? (You don't have any of them.)

How do we show we have zero of something? How do we write zero? (0.)

Students may not yet fully understand that in the counting order each number is 1 more than the number said after, but this is an opportunity to share an example that helps lead them to this insight. They revisit this idea, as well as exploring the concept of more and less, in later lessons.

SMP2

Listen as students share their understanding of 0. Emphasize that numbers represent how many of an item we might have or count, and 0 means we have none.

Extension: Talk about this question or draw a response: "What would you like to have 0 of?" (e.g., monsters, spiders, lima beans, angry bears, stinky shoes).

Through their responses to this question, you can assess students' understanding of 0. They should be giving examples of things they do not want to have.

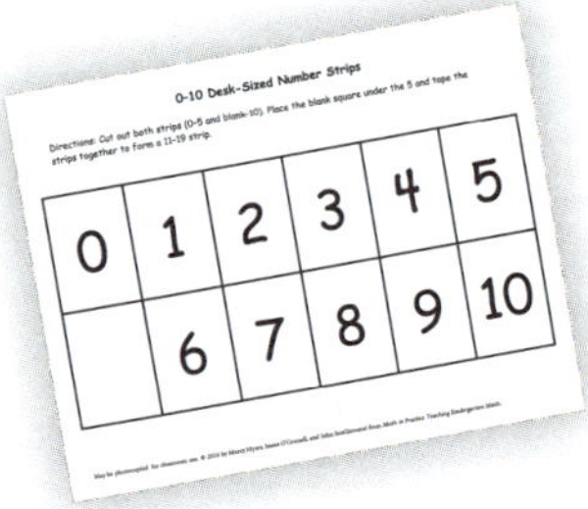

Number Towers

Students build towers with number cubes and place them on a number strip to show that each number is 1 more than the previous number.

Materials: number strip, 55 connecting cubes

Have students work with partners to build towers for each number 1–6 and place it on the number strip.

As students work, ask questions to help them make connections to the concept of "each number is 1 more."

What do you notice about the towers as you build the numbers?

After students have towers for 1–6, ask them to describe how their towers look (getting bigger every time; 1 more cube every time).

> This lesson could be done in a small group or with partners due to the number of cubes needed to build the number towers.

Ask them to predict what the next tower will look like.

Make a tower of 7 cubes and see if you are right.

Is the tower for 7 cubes 1 taller than the tower for 6 cubes?

Ask them to predict what the next tower will look like.

Make a tower of 8 cubes and see if you are right.

Is the tower for 8 cubes 1 taller than the tower for 7 cubes?

What do you think the towers for 9 and 10 will look like?

Have students build towers of 9 and 10 to check their predictions.

Let's count each number tower starting at 1.

What happened to the numbers as we counted?

Ten Black Dots

Black dots that count to 10 are shown as interesting illustrations in *Ten Black Dots* by Donald Crews.

Before Reading:

Draw 1 black dot on the board.

What could 1 black dot be?

Draw 5 black dots in a scattered arrangement.

What could 5 black dots be?

Draw 10 black dots in a line.

What could 10 black dots be?

Then tell the students to watch for what the author makes from 10 black dots.

During Reading:

Read *Ten Black Dots* by Donald Crews.

Encourage students to make predictions while you read the book.

How many black dots will be in the next picture?

After Reading:

Look at the diagram with 1–10 dots at the end of the book.

When you look at these dots, what do you notice? (Each number shows 1 more dot.)

Revisit the illustrations and brainstorm other ideas for 1 to 10 black dots.

Students then create their own pictures using 1 to 10 dot stickers.

Students can write at the bottom of the page "___ dots can make ________."

Pages can be put together to form a class book.

Additional Ideas for Support and Practice

The following ideas extend students' skills at counting groups of objects to 10 and subitizing and provide meaningful practice with these skills.

Subitizing activities are included here. Repeated experiences visualizing numbers 1–10 allow students to immediately recognize the numbers without counting. These experiences should be a part of early number work.

Counting Page by Page

As numbers are introduced, students create pages to a counting book. The pages include:

- the number
- a ten frame they fill in to show the number (or 2 ten frames that show the number in two different ways)
- a square that becomes a dot card to show the number
- a drawing to show a quantity of anything the students would like to illustrate

By this time some students are subitizing, simply knowing what number is represented, rather than having to count each object.

The pages are saved and stapled together to form a book later in the year.

Ten Frame Match Center

Students practice their counting (or subitizing) skills by matching ten frame cards to the corresponding number cards.

Common Error

Students may have trouble differentiating between 6 and 9 on number cards. It is helpful to underline the numbers so students see how the number should be positioned with the line at the bottom.

Each pair needs a set of 0–10 ten frame cards and a set of 0–10 number cards.

Students lay the number cards in front of them in a row 0–10.

Students put the ten frame cards in a pile facedown.

Partners take turns picking a ten frame card and finding the number that shows how many dots are on the card.

When all of the cards are matched, players start again.

Students practice counting (or subitizing) skills by matching mini ten frames to the corresponding number cards.

Count the Stars

Students practice counting in scattered configurations by using counters on a template with stars.

In small group, give each student a Count the Stars sheet and 10 cubes or counters.

How can we use these manipulatives to count the number of stars on the page?

Students may suggest placing 1 counter on each star and then counting to find the total.

Observe students as they count.

> Students should be naming only one number as they touch a cube for one-to-one correspondence.

Click, Clack, Splish, Splash

Duck, Farmer Brown, and the animals count to 10 and back.

Before Reading:

Have the students chorally count to 10.

During Reading:

Read *Click, Clack, Splish, Splash* by Doreen Cronin.

As students hear the story, have them predict.

What number will be next?

How do you know?

After Reading:

Students choose a number 1–10 and draw a picture of the animals.

Put the pictures together to make a class book that counts forward and back.

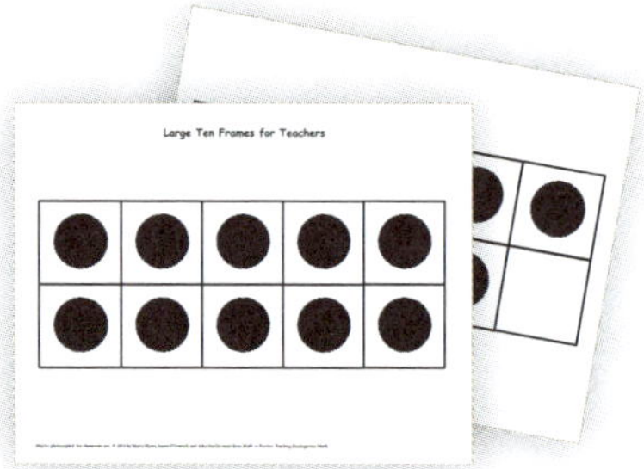

Fast Flash

Students practice to be able to automatically recognize the number of dots shown without having to count each dot (subitize).

Flash a ten frame showing a number of dots for about 3–5 seconds, and then turn it over.

How many dots did you see?

How did you know it was that number?

Continue flashing numbers in random order.

As students begin to identify the numbers, gradually decrease the time to about three seconds.

As an option, students can show the number they saw by using their fingers. All students have an opportunity to respond to the question and the teacher has an opportunity to assess students' ability to subitize.

Match My Number

Students learn to quickly recognize numbers on a ten frame and then use counters to build the numbers on their own ten frame.

Flash a number on a ten frame.

What number did you see?

How can you show this on your ten frame?

The activity can be extended with nontraditional arrangements of numbers.

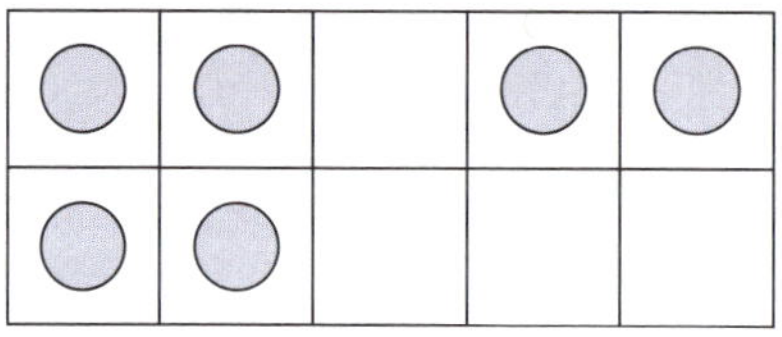

Students could play this as a partner game with one player flashing a ten frame card and the other player making the number on a ten frame.

Model how one partner will hold up a ten frame card for three seconds and then place the card facedown.

When the card is placed down, the second partner will build the number they saw.

Differentiation

If students struggle to identify the number of dots, revisit the idea of how many are on one side of a ten frame, how many are on a whole ten frame, and practice with models of 1–5, then extend to models of 6–10.

Differentiation

If students struggle to identify the number, revisit the Building a Ten Frame activity.

Giving students a set three-second interval to view the card allows enough time for them to see the quantity, without giving them time to count each dot.

Show It Three Ways

Have students pick a 1–10 number card and show their number three different ways.

Students might show it on a number line, a ten frame, or make their own arrangements.

Then, they pick a new number and show that number three different ways.

This student uses cubes to show the number 6 on a ten frame and also shows it by covering 6 on a number line. She could also show it by making a train of 6 cubes, drawing a dot card with 6 dots, showing 6 a different way on a ten frame, or any other way to accurately represent 6.

Talk About It/Write About It

For more ideas on ways to generate math talk, see Math in Practice: A Guide for Teachers, Chapter 4.

Use the following as discussion starters or journal prompts:

In your math journal, draw three different ways to show the number 7 on ten frames.

Max said there were 8 cubes in the cup. How can you know if Max is right? Explain.

Molly got confused counting the cubes that were in a circle. What could she do so she wouldn't be confused?

RECOGNIZING AND WRITING NUMBERS TO 10

Students continue to count objects, but these lesson ideas include a written component in which students either select or write the number to match the total objects in the group.

Counting Bags of Cubes

In a small group, students count the number of counters in bags and match the numeral to the set.

Materials: bags with a different number of counters (6–10) in each bag, number cards to match each bag

Give each student a bag of counters.

You each have a bag with a different number of counters in it.

How can we figure out how many counters are in each bag?

Go ahead and count what's in your bag. Be ready to share.

Students may slide each counter as they count to keep track or count in an organized manner.

As students are counting, watch for one-to-one correspondence as they touch each counter.

How did you count your cubes?

Why did you line up the cubes?

Why did you touch each cube when you counted?

How did you know that there were 7 cubes in your group?

Summarize the key ideas about counting. "I saw some of you lining up the cubes before you counted so you knew where to start. I saw some of you touch each item as you said each number so you were sure to count each cube. I noticed that you knew that the last number you said when you were counting was how many cubes were in that group. You used a number card to label the number of cubes, to show how many cubes were in your bag."

Show students number cards 6–10 and have them select the card to match their group of cubes.

How many were in your bag?

Did anyone else have the same number as you?

Switch the bags and repeat the process several times.

FORMATIVE ASSESSMENT

Version 1

Give each student a set of number cards for 0–10.

Ask the student to lay the cards down in order, faceup.

Show a ten frame card and ask students to hold up the number that shows how many dots are on the frame.

Students should be able to match the number card to the ten frame shown.

Record observational notes of student responses.

Version 2

Give the student a set of ten frame cards and number cards for 4, 7, and 9 in random order.

Tell the student to match the number card to the ten frame card.

The student should be able to match each number to the correct frame.

Some students may need to count the dots on the ten frame card to match the number. With continued practice, they should get quicker with subitizing.

If working with 0–10 cards is too cumbersome for students, numbers 1–5 and 6–10 could be assessed separately.

Use the observational notes page.

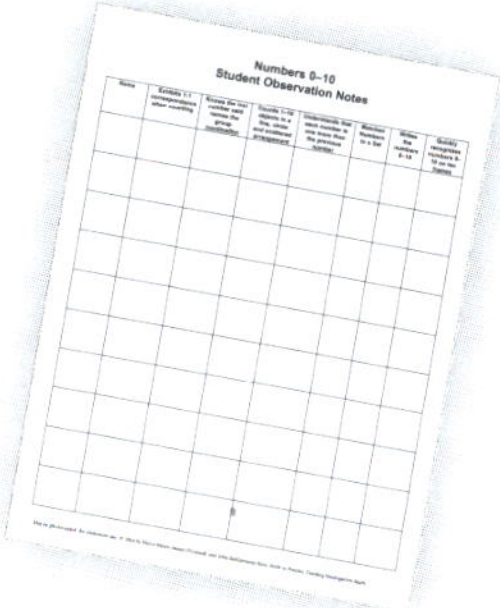

Writing Numbers 0–10

While searching for numbers throughout the school, take pictures and then have students practice writing numbers 0–10.

Walk through the school searching for numbers 0–10.

Take pictures of each number you find.

Go back to the classroom and review the pictures by projecting them on the board.

Give each student a dry erase board and marker.

What number did we find on the cafeteria door (e.g., 10)*?*
Show how to write the number 10 on your board.
Hold up your board to show the class what you wrote.

For students who wrote 10 as 01, ask

Does your 10 look like the 10 in the picture?
What number comes first when we write 10?
Write the 1. Then write the 0.

Continue with other pictures and numbers found in the school.

Numbers could be strategically placed throughout the school ahead of time if necessary.

Differentiation

Some students may benefit from using a number strip to copy the number.

As students write numbers, make note of students who write the numbers backward. They may need a number strip to practice copying numbers correctly. Ask students, "Does your 4 look like this 4?"

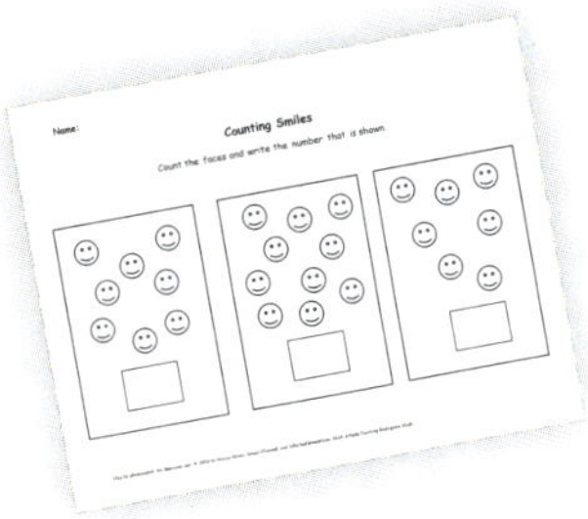

FORMATIVE ASSESSMENT

Have students count smiley faces in scattered arrangements and then write the number.

Counting Arrangements Scatter 2

Count the faces and write the number that is shown.

This student used the strategy of drawing a line through each face to count and accurately write the number for each group.

Counting Arrangements Scatter 2

Count the faces and write the number that is shown.

This student did not mark each face and miscounted the groups of 8 and 7 faces. Her count was off by 1 for each number indicating that she may have counted a face twice.

It would be helpful to observe the student as she counts to determine if the problem is that she is confused by the configurations. If so, she would benefit from practice marking each item as she counts.

Additional Ideas for Support and Practice

The following ideas extend students' skills at recognizing and writing numbers and provide meaningful practice.

Matching Numbers to Sets

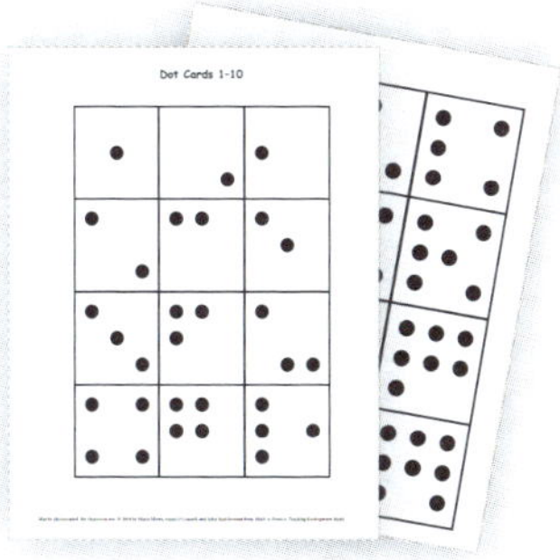

Students match dot cards and number cards.

Prepare dot cards and number cards for numbers 0–10.

Students sort the dot cards into sets and label with the number cards.

Cups of Counters

Players work with partners to count and model sets of counters to 10.

Player A puts any number of counters (maximum 10) in a cup.

Player B spills the cup and puts the counters on the ten frame and tells how many there are.

Both players draw a picture of the ten frame and counters and write the number.

Then players switch roles and repeat.

Counting Collections Center

Students practice counting collections of various items and write the number of each item.

Place collections of 1–10 objects in bags, cups, or trays in the center area for students to count. Students label the collection by writing the number on sticky notes and sticking it to the collections.

Additional ideas for collections are pennies, beads, erasers, crayons, paper clips, and pom-poms.

Climb to the Top

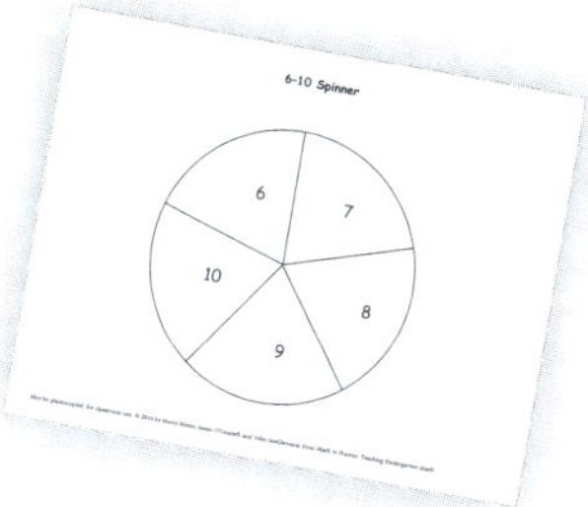

Students spin a 6–10 spinner and then write the number to see which number wins.

Materials: 6–10 spinner and Climb to the Top recording sheet

Before beginning, students predict which number will make it to the top first.

Students then spin a number and write the number spun in the box on the recording sheet beginning at the bottom of the column.

Students continue spinning and writing the number until all of the spaces for a number are filled; that number reached the top first. They color the happy face to show it was the winner!

Vocabulary

arrangement
counting order
one more
ten frame
zero

ONLINE RESOURCES

General Resources

Additional Problems
Dot Cards 0–10
Five Frame Mats
Large Ten Frames for Teachers
Number Cards 0–10
Number Strip
Numbers 0–10 Student Observation Notes
Teacher Ten Frame (blank)
Ten Frame Cards
Ten Frame Mats

Resources for Specific Activities

6–10 Spinner
Climb to the Top 6–10 Recording Sheet
Count the Stars
Counting Smiles Recording Sheet

These resources are available at http://hein.pub/MathinPractice, keycode MIPGK.

Domino Dot Numbers

Students use dominoes to practice counting.

Try the following domino activities for recognizing and writing numbers:

- Pick a domino, count the total dots, match the number card.
- Pick a domino, count the total dots, write the number.
- Pick a number card and then find dominoes with the number of dots that match the number on the card.

Counting Soup

Students scoop pom-poms or cotton balls from a bowl and then count and label the amount.

Materials: large plastic bowl; 10 large pom-poms or cotton balls; a large ladle, spoon, or scoop; paper plates, sticky notes; and pencils for each student or pair of students

Students spoon out pom-poms onto the plate and then count.

They write the number of pom-poms on the sticky note to show the total and stick it to the plate.

Student can repeat or take turns with a partner.

MODULE 3

Counting and Cardinality and Place Value *Numbers 0–20*

About the Math K.CC.3; K.CC.4; K.CC.5; K.NBT.1

Students continue to explore counting and cardinality to 20.

The key ideas focused on in this module include:

- counting and naming the number of objects in a group of 20 or fewer
- recognizing and writing the written numerals to 20
- recognizing that 11–19 are 10 and 1, 2, 3, 4, 5, 6, 7, 8, or 9 more.

Gradually, as we extend the numbers they are counting, students see that the big ideas about counting and numbers remain the same.

- A number of objects is not changed by the configuration.
- In the counting order, each number is 1 more than the previous number.
- When counting, the last number said tells how many objects are in a group.

These big ideas are central to the study of numbers in kindergarten, and these ideas form the foundation for students' further explorations with numbers.

Students continue to develop their skills by counting larger groups of objects, using one-to-one correspondence when counting objects in different configurations such as lines, arrays, and circles. Now, their reading and writing of numbers extends to 2-digit numbers to 20. They begin to notice patterns in numbers as they see the 0–9 sequence repeat as they count 10–19 (see Figure 3.1). As they write these 2-digit numbers, they are not only challenged to form each numeral correctly but also to be cognizant of the order of the digits they write (e.g., 16 being 16 and not 61). Some may also still express confusion with number names like *eleven* and *twelve* that do not follow the name patterns of the other teen numbers.

Students do not need to count scattered configurations with more than 10 items.

1	2	3	4	5	6	7	8	9	10
11	12	13	14	15	16	17	18	19	20

Figure 3.1 Students begin to see the patterns in how counting numbers are represented.

Understanding the numbers 11–19 as *10 and some more* is a foundational understanding that leads to the place value concepts that students explore in grade 1. When counting objects from 11–19, students are merging their counting skills with a new place value understanding. Students use manipulatives to practice counting and also to see how the teen numbers connect to 10, making discoveries about those numbers. Experiences in which students see 11–19 as 10 and some more, like trying to place 13 counters on a ten frame, will help them gain insight into the quantity represented by these numbers (see Figure 3.2).

With the transition to counting 2-digit numbers, counting becomes more complex for some students. The number words can be confusing and more objects to count adds challenge to the counting process. Students need opportunities to gain familiarity and confidence with this new set of numbers.

Figure 3.2 When attempting to show 13 counters using a ten frame, students realize that 13 is more than 10, that it is 10 and 3 more.

Learning Goals

I can count 20 or fewer objects no matter how they are arranged.

I can write the numbers 11–20 to show how many are in a group of 11–20 objects.

I know that when I count, every number I say is 1 more than the number before it.

I know that the numbers 11–19 show 10 and some more.

The hands-on activities in this module build on those in the previous modules. Now students count 20 or fewer objects, recognize and write numbers 0–20, and begin to explore place value concepts. They continue to use cubes and counters and now transition to using double ten frames to show the numbers 11–20. Exploring numbers with varied manipulatives allows students to recognize that 13 is 13 regardless of how it is represented, and that it always can be broken into 10 and 3 more (see Figure 3.3).

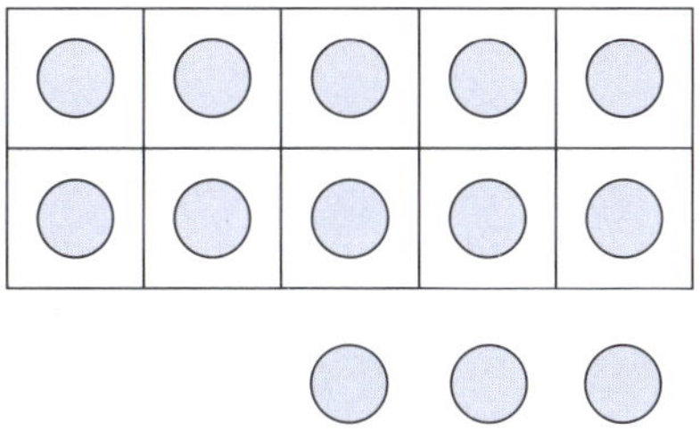

Figure 3.3 Students visualize 1–20 with counters, double ten frames, and connecting cubes (linked as 10 and some more) and continue to observe, discuss, and extend their understanding of critical number concepts.

Exploring the Progression

PREVIOUS		NOW		NEXT
Preschool		**Grade K**		**Grade 1**
One-to-one correspondence for 1–5; possibly rote counting to 20 but not necessarily reading or writing the numerals 11–20	»	Counting 20 or fewer objects; reading and writing numbers to 20; understanding that teen numbers are 10 plus some more	»	Reading, writing, and counting numbers to 120; exploring concepts of place value with 2-digit numbers

Ideas for Instruction and Assessment

EXPLORING NUMBERS TO 20

Rather than separating counting, writing, and place value concepts, the following activities integrate these three aspects of numbers to 20. Students are introduced to the numbers 11–20 as they count, examine their relationship to 10, and explore how to write the values.

Introducing Numbers 11–20

Students explore numbers 11–20 with actions, objects, drawings, and discussions.

Give each pair a bag with 11 cubes in it.

Work with your partner to decide how many cubes you have.

Write the number 11 on the board.

What do you notice about the number 11?

Lessons in This Module

Exploring Numbers to 20

Introducing Numbers 11–20
Counting Arrangements
Which Number?
20 Big Trucks in the Middle of the Street
10 and Some More on a Ten Frame
May I Have Some More?
Double Ten Frames
Towers of 10 and Some More (Observing the Digits in 11–19)
Number Tents

Students have had prior experience with foundational counting skills. As they are introduced to the numbers 11–19, they explore the written numerals as well as the new look of these teen numbers. They discover that the values are 10 and some more and that the numerals are now 2-digit. These big ideas are integrated throughout the lessons.

- *Do they need to review that the configuration does not change the quantity?*
- *Do they need to practice one-to-one counting?*
- *Do they understand that the last number named says the quantity in the group?*

If students are unable to count objects with one-to-one correspondence and cannot count quantities to 10, continue to focus on the concepts in Module 2.

It is not necessary that students immediately understand this distinction, but that they are introduced to the term. Continue to remind them of the meaning of digit as you explore 2-digit numbers.

Record the term *digit* on the word wall or Math Talk chart. Tell students that *digit* is a word we use when we talk about numbers. The number is 11, but the digits are 1 and 1.

What other numbers with 2 digits have we explored?

Put a basket of pattern blocks at each table.

Show me 11 yellow hexagons.

Show me 11 red trapezoids.

Show me 11 green triangles.

Show me 11 orange squares.

This lesson is a variation of the lessons used in modules 1 and 2 to introduce numbers. Modify it in any way that meets your students' needs. The important thing is to take the time to explore each number and allow students to count objects and discuss the new notation. Repeat a similar lesson for each number introduced.

Have students stand up.

Hop 11 times.

Do 11 jumping jacks.

Give each of the students a paper and have them draw 11 of something.

Have them label their drawing with the number 11.

Have them share their drawing with the class.

Counting Arrangements

Students count cubes in different configurations.

The scatter configuration is not required for numbers greater than 10.

Place a container of cubes in the center of the table.

Place 12 cubes in a row in the center of your small-group table. Do not tell the students that it is 12 cubes.

Ask a student to touch and count the cubes and tell you how many there are.

Change the configuration to an array, but don't add any cubes.

Have a student count and tell you how many.

Is it still 12 cubes?

Did anyone know how many cubes there were without even counting them?

Change the configuration to a circle, but don't add any cubes.

Have a student count and tell you how many.

Observe the student to see how he organizes to be sure the count is correct (e.g., touching and sliding as he counts or finding a way to remember where he started in the circle).

Was it still 12 cubes? How do you know?

What might you do to be sure you count each cube just once?

Did anyone know how many cubes there were without counting them?

Have each pair take 11 cubes from the container in the center of the table. Have partners check to be sure they have 11.

Ask partners to work together to arrange the cubes any way they'd like (e.g., horizontal line, circle, array, diagonal line).

Look at the other pairs' cubes. Do you notice any that are arranged like yours?

Do you notice any that are arranged different from yours?

How many cubes do you have?

What do you notice about all of the cubes?

How can they all be 11 when they look different?

Does the number of cubes change when you arrange them different ways?

Repeat the activity and questions by asking students to take 13 cubes from the container.

Emphasize that 13 cubes is 13 cubes no matter how they are arranged.

Give students a copy of the How Many Dots? student sheet.

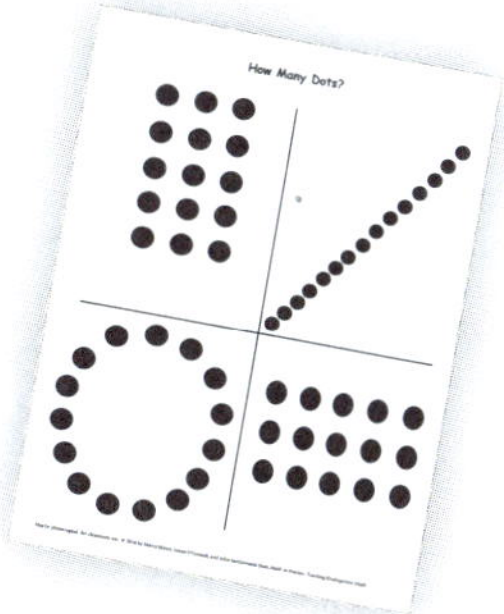

Have them count the circles in each section and tell their partner how many are in each circle.

Do these all show 15 circles?

But they don't look the same. Is that okay?

Have students take 16 counters.

Show 16 any way you'd like.

Watch as students work. Have them share their arrangements.

Ask students to count aloud to check for 16.

Observe as students count to be sure they are showing one-to-one correspondence, saying each number without skipping any, and showing strategies for finding a place to begin and end their counts.

Which Number?

Students count collections of manipulatives for 11–20 and label the set with number cards.

Prepare bags with various sets of 11–20 manipulatives and number cards for each pair of students.

Dump out one bag of counters on the table and ask a student to count them out loud.

The other students can whisper count as the student touches each counter.

How many counters are there?

How do you know there are 13?

Give each pair a set of 11–20 cards.

Talk with your partner and decide which card shows 13.

When you have decided, hold up the card so I can see it.

Common Error

Students may have trouble with the names of some numbers in the 11–20 sequence. 11, 12, 13, and 15 do not restate the single-digit number names like 14, 16, 17, 18, and 19. 20 begins a new decade and counting pattern.

Repeat with another set of counters.

Give each pair of students several bags of counters with 11–20 objects.

Count them and place the number by each set.

Watch as students count and select the card to match each set of counters.

After students have counted and selected numbers for several bags, have them tell how many counters were in their bags.

Were any numbers easier to say than others?

How did you know which number to put with each set?

What do you notice about these numbers? (They all have a 1 in front of a familiar number 1–9 except for 20—that has a 2.)

Have sets of partners switch bags to practice counting sets and selecting the appropriate numbers 11–20 to label their sets.

Use the observational notes page.

FORMATIVE ASSESSMENT

Put 20 teddy bears on the table.

Ask the student to count out 14 teddy bears.

Record observational notes of student work.

Return the 14 bears to the set of 20.

Ask the student to count out 17 teddy bears.

Continue to record notes of student work.

If the student is not able to produce a set of bears by counting when given a number, then put out 14 bears in a line arrangement and tell the student to count the bears.

Repeat with a line of 17 bears.

Check to be sure students are able to use their counting skills (e.g., knowledge of the rote counting sequence, one-to-one correspondence, understanding that the last number named is the quantity) to count a line of bears. If any of those number concepts is deficient, address them in small-group instruction.

20 Big Trucks in the Middle of the Street

SHOW IT!

Students listen to a story, *20 Big Trucks in the Middle of the Street* by Mark Lee, that counts to 20 and show each number on a double ten frame.

Materials: a double ten frame and 20 counters for each student

Before Reading:

Introduce the book.

What are different types of trucks that you might see on the road?

Tell the students that you are going to read a story about 20 big trucks and you will count each truck as you read about it.

Give each student a double ten frame.

What do you notice about this frame? Does it remind you of anything?
Count the spaces on the 2 frames: How many are there?
We call this a double ten frame. Does that make sense? Why?
We are going to use this double ten frame to show trucks in the story.

Give each student 20 counters.

During Reading:

Read *20 Big Trucks in the Middle of the Street.*

As a new truck is named, students place a counter on the double ten frame to count the trucks.

Students should fill the ten frames left to right and top to bottom to show a traditional counting arrangement.

When the fifth counter is added, ask

What do you notice about the number on the double ten frame?

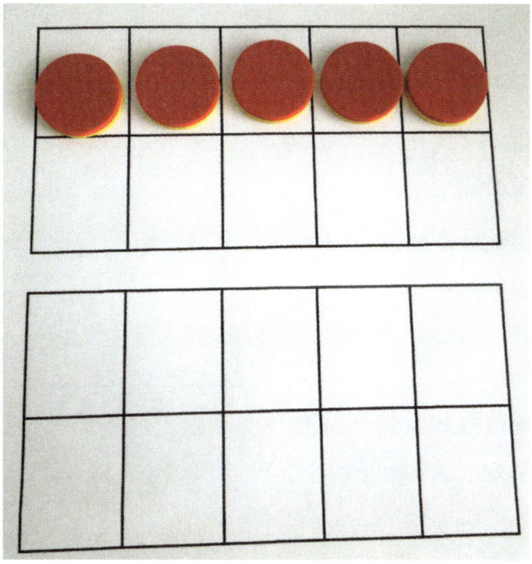

What number will come next?
Let's read to see what type of truck is number 6.

Students continue to add counters for each truck.

When all 20 trucks are shown in the middle of the street, count the trucks in the illustration.

In the story, some pages count several trucks. Be sure to check that students are adding the correct number of counters.

How many trucks did we count?
How many counters do we have on the double ten frame?

Let's read to find out what happens to the trucks.

Keep the counters on the double ten frame.

Continue reading.

At the end of the story, only the ice-cream truck remains, but students should still have 20 counters on their frames.

After Reading:

How many trucks did we count?

Write the number *20* on the board for students to see.

How did we show 20?

Look at the double ten frame. What do you notice?

How many ten frames did we fill?

If we have a double ten frame that is full, what number does that show?

Let's count to 20 as we take the counters off of the double ten frames to prove there are 20.

Thinking Through a Lesson

10 and Some More on a Ten Frame SHOW IT!

In a small group, students begin to learn about place value by using counters to show numbers 11–19 on a ten frame as 10 and more ones.

Materials: number cards 11–19, bag or bowl of 19 counters, and a ten frame for each student in the small group

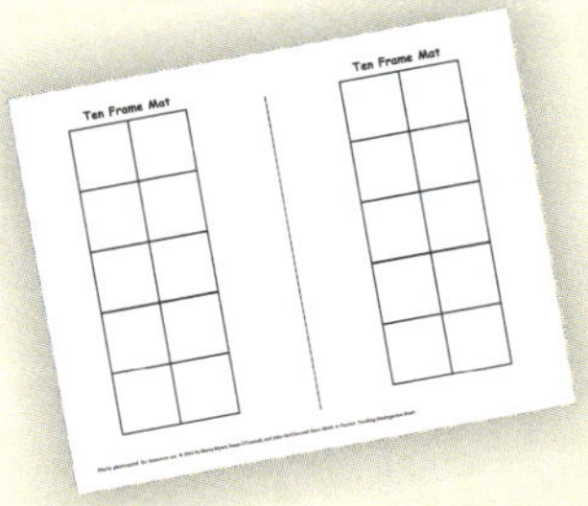

Give each student a bag of counters and ask them to take out 11 counters.

Choose a number card to show the number of counters you have.

After students put the number card with 11 near the counters, give each a ten frame.

What do you remember about ten frames? (They have 10 spaces. *Or,* We use them to show numbers; you can put 10 counters on them.)

Put your 11 counters on your ten frame.

Rather than simply giving students 11 counters, have them count out 11 from a bag of counters. Not only do they get counting practice, but they have now identified that they are working with 11 counters and use that understanding to make sense of what happens when they place them on the ten frame. Observe that students have one-to-one correspondence and do not skip numbers as they count.

SMP4

Students should be proficient with numbers to 10 before moving on to teen numbers. If a student does not know that a ten frame shows a group of 10, revisit numbers to 10.

Observe as students try to fill their ten frame.

What happened when you put your counters on the ten frame? (We couldn't put them all on. *Or,* They didn't fit.)

Why didn't they fit? (There were too many. *Or,* There is only room for 10.)

What do you mean there is only room for 10? (There are only 10 spaces so only 10 will fit.)

How many wouldn't fit? How many were left over? (1.)

What did you do with the extra counter? (I just put it out here.)

So could we say 11 is 10 and 1 more? (Yes.)

SMP3

Observe students to see what they do with the 1 extra counter. Do they try to squeeze it onto the ten frame even though there are only 10 spaces? Don't tell them how to deal with it. Observe to see what they do.

On the board, draw a model of the 10 counters on the ten frame and the 1 leftover counter.

Write: *11 is 10 and 1 more.*

Where is the 10? Touch it to show me.

Where is the 1 more? Touch it to show me.

SMP2

Students benefit from repeated practice counting and identifying numbers.

Have students remove their counters from the ten frame and remove the number card label.

Count out 12 counters.

Choose a number card to show the number of counters you have.

Put your counters on your ten frame.

Observe as students fill their ten frames.

What is happening? (The frame is full again. *Or,* There are extra counters that don't fit on the frame.)

How many are left over? How many don't fit? (2.)

So, is 12 more than 10? (Yes.)

How many more than 10? (2.)

How do you know that? (There are 2 counters that don't fit.)

Having the number card by the ten frame model allows students to see both the number and the visual model. As they see the 2 extra counters, they may notice the 2 in the number 12.

Draw a ten frame on the board and record: *12 is 10 and 2 more.*

Have students put the counters back in the bag.

Ask them to count out 13 counters and label them with the correct number card.

Turn and tell your partner: Do you think you will be able to fit 13 counters on your frame? Tell your partner why you think that.

Listen to see which students immediately recognize that 13 is more than 10. See if any interject that it is 3 more. Are they beginning to notice a pattern?

Have students try it.

Were you right? Did they all fit? (No.)

How many were left over? (3.)

So what is 13? 10 and how many more? (3 more.)

Draw the ten frame on the board and label it as *13 is 10 and 3 more.*

Have students put the counters back in the bag.

Have them look at the models and statements on the board.

SMP7, SMP8

Read the phrases together and ask students what they notice.

11 is 10 and 1 more.

12 is 10 and 2 more.

13 is 10 and 3 more.

We are gathering and recording data to help students see the repetition that is occurring. Rather than telling them to look at the 4 in 14 to see how many are left over, we want them to discover this on their own, making sense of these teen numbers. We create models, record the data, ask them to think about the data, ask them to predict for the next example, and continue to ask them to justify their thinking to push them toward this discovery.

Allow any students who may have insights to share their ideas.

Then, ask students to count out 14 counters and label them with the correct number card.

Turn and ask your partner: Do you think you will be able to fit 14 counters on your frame? If not, tell them how many more you think you will have?

Why do you think that? (Because it's 14 and I think it is 4 more than 10 because 13 was 10 and 3 more.)

Try it.

Did you have 4 more? (Yes.)

Have students continue to investigate with the numbers 15–19, predicting how many will be left over, showing them on the ten frame, and saying it as *10 and __ more.*

Continue to record the *10 and __ more* phrases on the board.

Throughout the process, have students share their insights with partners and the class.

As students look at the numbers and talk about what they see, they begin to notice that the number of leftovers matches one of the digits in 11 or 12 or 13 or 14. They may not all truly understand the connection now, but they are beginning to see repetition that gives them clues as to how these numbers are formed.

Showing teen numbers as 10 and some more is the foundation of place value. This student shows the number 16 as 10 and 6 more.

May I Have Some More?

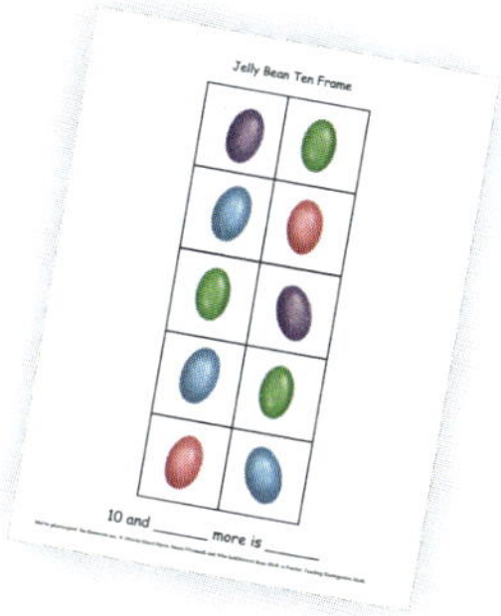

Students use jelly beans to show numbers 11–19 on a ten frame as 10 and more ones and label a number to match.

Materials: laminated ten frames with images of 10 jelly beans and the words "10 and ___ more is ___" (or pages placed in page protectors so they can be written on); dry erase markers

Distribute one jelly bean ten frame to each student.

Here are the jelly beans I gave you.

How many jelly beans do you have? (10.)

Well, you happen to love jelly beans, so you asked politely if you could have more.

Of course I agreed and gave you 1 more.

Draw the jelly bean I gave you beside the ten frame.

> *Observe students when you ask this question. Which students needed to count to answer the question? Which students knew that there were 10 jelly beans because the ten frame was full?*

Distribute a dry erase marker to each student and observe as they draw a jelly bean to the side of their ten frame.

How many jelly beans do you have now?

How do you know there are 11 beans on your jelly bean mat?

Look at the bottom of your mat.

What does it say? (10 and _____ more is _____.)

What could you write in the blanks?

Tell your partner what you wrote.

Let's read what we wrote.

Touch each part as we read it: 10 and 1 more is 11.

That is super, but it is not where your story ends!

> *Observe to make sure the numbers are going in the correct blanks on the mats.*

Have students erase the numbers 1 and 11 from "10 and __ more is __" on their mat.

You are very happy with your 11 jelly beans, but you love them so much that you want more!

So again, you ask politely if you could have more and, of course, I agree and give you 1 more jelly bean.

Draw the next jelly bean I gave you.

How many jelly beans do you have now?

How do you know there are 12 beans on your jelly bean mat?

Do I need to go back and count the ten beans on the ten frame by ones?

Look at the bottom of your mat.

Let's use numbers and words to explain what we see on our jelly bean mats.

Fill in the blanks.

Let's read what we wrote.

Use your finger as we read what you wrote: 10 and 2 more is 12.

That is super, but it is not where your story ends!

Continue adding one jelly bean at a time and filling in the blanks on the mat to coincide with the jelly beans shown and discussing what students see as they build each new number.

Double Ten Frames SHOW IT!

Students show numbers 11–19 on double ten frames.

Materials: ten frame, double ten frame, counters, number cards

Give students counters and a ten frame.

Show me 15.

How many counters are on your ten frame?

How do you know?

How many counters are off the ten frame?

How do you know?

Could we find a way so you wouldn't have to count them?

Turn and talk with your partner about how you could show them without recounting them.

Have student share ideas.

What if we had another ten frame?

Give students a double ten frame and take back the single ten frame.

Put your counters on your new ten frames.

This is called a double ten frame.

Can you see the number better now?

How do you know it is 5?

Did you have to count them?

Can you see 15 is 10 and 5 more?

Differentiation

If needed, remind students that if 1 side of their frame is filled it shows 5 because there are 5 on each side of their frame.

Have students clear the counters from their ten frames.

Show 12 on your double ten frame.

How do you see 12? Explain.

12 is 10 and how many more?

Have students pick number cards (11–19) and show the numbers on their double ten frames.

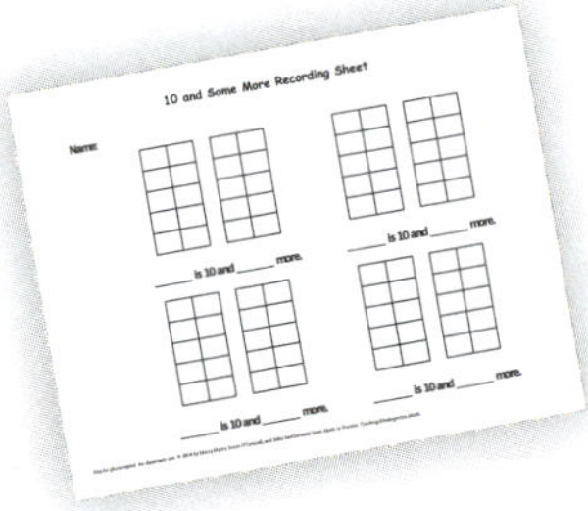

Give each student a 10 and Some More recording sheet and have them record the number on the double ten frames and fill in the phrase __ *is 10 and* __ *more* for each frame.

Multiple representations allow students to see numbers and make connections between them. This student shows 13 using a full ten frame and 3 more ones on a second frame. He matches a train of 10 linking cubes and a train of 3 more cubes to also show 13.

FORMATIVE ASSESSMENT

Give a student 15 counters.

Put the counters on the double ten frame to show how many there are.

Observe as the student places the counters.

How does 15 show a group of 10?

The student should be able to say that 15 is a group of 10 and 5 more ones.

Continue with another teen number.

If the student is not able to say that 15 is 10 and 5 more, prompt her with questions.

Where is a group of 10?

How many do you see on the other frame?

How many more than 10 do you see?

Provide additional opportunities to split teen numbers into 10 and some more for those students who are not able to verbalize the 10 and some more ideas.

Record student response on the observational notes page.

Towers of 10 and Some More (Observing the Digits in 11–19) SHOW IT!

Students use connecting cubes to show numbers 11–19 as a tower of 10 and some more.

Students need 20 connecting cubes each.

Count out 11 cubes.

Let's use these cubes to make a tower of 10 cubes and see what is left.

How many towers of 10 do you have?

How many leftovers?

You have 1 tower of 10 and 1 left over. That is 11.

Draw a model of a tower and 1 more on the board and write *11* by it as shown below.

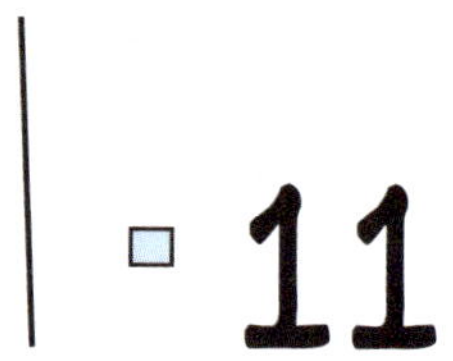

Have students break the trains and begin again with a new number.

Count out 13 cubes.

Let's make a tower of 10 cubes and see what is left.

How many towers of 10 do you have?

How many leftovers?

You have 1 tower of 10 and 3 left over. That is 13.

Draw a model of a tower and 3 more on the board and write *13* by it.

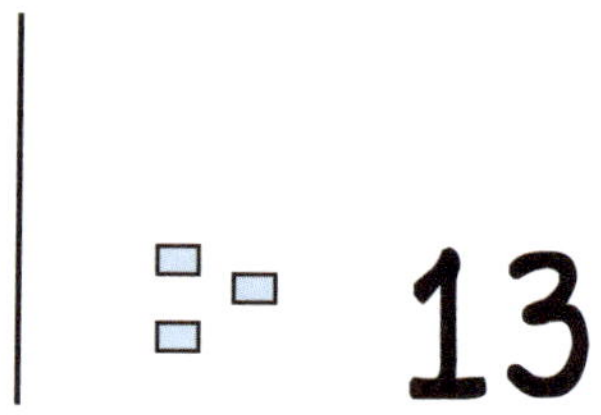

As students are exploring the numbers, they begin to see the connection between the meaning of each digit and the 10 and some more idea. In 15 there is one 10 and 5 more and it is written as 15. Although we don't refer to them as tens and ones digits in kindergarten, students are beginning to notice foundational place value concepts that they will explore further in grade 1. Right now, this insight may help some students realize why the 1 is written before the 5 as they translate the meaning to 10 and 5 ones.

Try it with another number following the same process.

What do you notice about the numbers and the models? (They show a tower of 10 and some leftovers.)

What if we have 1 tower of 10 cubes and 5 more?

What number do you think we'd have? Why?

Try it to see if you are right. Show the number 15.

Allow students time to work.

Have students use their cubes to build towers of 10 and some more for numbers 11–19.

What do we know about showing teen numbers with a group of 10?

Draw pictures in your journal to show teen numbers.

FORMATIVE ASSESSMENT

Have students draw 11 (or 12 or 13 . . .) of any item they'd like and then show you 10 and __ more.

Draw a picture to show the number _____.

Show it as 10 and some ones.

Students should be able to accurately show that number of objects and then show the number as 10 and some more remaining ones.

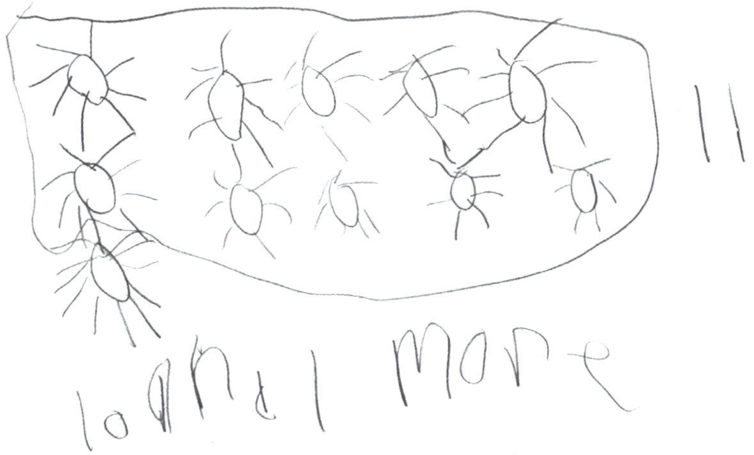

This student shows 11 suns as a group of 10 and 1 more sun. She used an array (2 rows of 5 like in a ten frame) to draw the 10 suns and then circle them. She records 10 and 1 more to show her understanding.

Number Tents SHOW IT!

In a small group, students create teen numbers using number tents, build the number using connecting cubes, and write the teen numbers.

Materials: number tents, 20 connecting cubes in two colors (10 of each color), sticky notes, a pencil

Use a set of number tents to show the number 14.

What do you know about 14? (It is after 13. *Or,* It is a teen number. *Or,* It is a group of 10 and 4 more.)

Slide the 4 tent off of the 10 tent to show students that the number tents have a group of 10 and 4 more.

These number tents show 14 as a group of 10 and 4 more.

Put the 4 back over the 0 on the tent.

Give each student 20 cubes in two colors.

Use your cubes to show 14.

Watch as students show a group of 10 and 4 more.

How could we place the cubes next to the tents to show 14?

Place a student's cubes near the tents.

Give each student a sticky note and a pencil.

We need to write the number 14.

Where could we look to help us write the number 14?

Write the number 14 on your sticky note and place it next to your cubes.

Observe as students write the number 14.

How does the number that you wrote match the number tent?

When you write 14, the 1 shows the one group of 10 cubes and 4 shows the 4 more.

Common Error

Students may write 41. Ask them if their number matches the tent showing 14. Ask which number they see first. (1.) Ask which number comes next. (4.)

Point to the tents.

The tents show 14 as a group of 1 ten and 4 more.

Give each student a set of number tents.

Use your tents and cubes to show the number 16.

Watch as students work and then ask,

How is 16 shown on the tents?

How is 16 shown with the cubes?

Give each student a sticky note.

Write the number 16 to match your tents and cubes.

Observe as students write.

How does the 16 you wrote match your tents and cubes?

It's your turn to show your own number!

Choose a number 11–19.

Show it with the tents and cubes and write the number to match.

Watch as students work.

Turn and talk to your neighbor about the number you made.

Make sure that your partner wrote the number correctly to match the tent.

Continue with other numbers to practice writing teen numbers.

The number tent shows 14. The linking cubes show a group of 10 and 4 more ones.

FORMATIVE ASSESSMENT

Option 1

Give students a piece of paper, folded into sixths, and a pencil.

Tell them you are going to say a number and they should write it in the box.

Say the number 14.

Students will write 14.

Tell students to write the next number in the next box. Each number goes in a box.

Continue with numbers 16, 19, 12, 13, and 11.

The numbers 11, 12, and 13 could be difficult for students to write because they do not begin with the words *one*, *two*, or *three*.

If students reverse numbers by writing the ones digit first and then the tens digit, reteach writing teen numbers and provide visual support for them to copy numbers as they practice writing them. Refer to teen numbers on a hundred chart or number strip to reinforce that the 1 comes before the other digit in all teen numbers.

This assessment can be done in small groups as each student writes the number given.

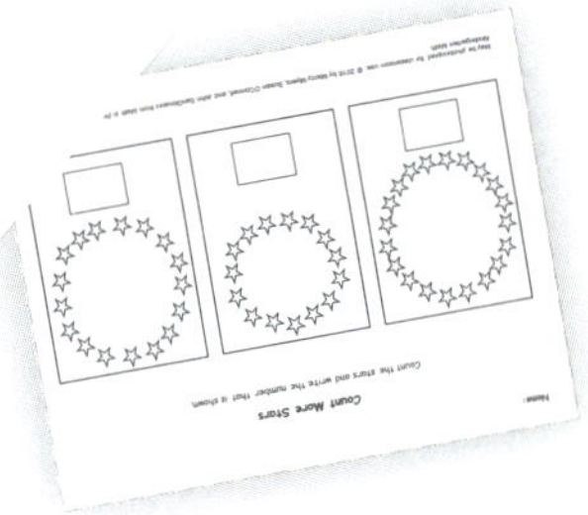

If students are not able to write the numbers, give them a number strip and have them point to the numbers to assess for number identification. Continue to provide practice time for writing the teen numbers.

Option 2

Have students count the number of stars in a circular configuration and write the number to label the group.

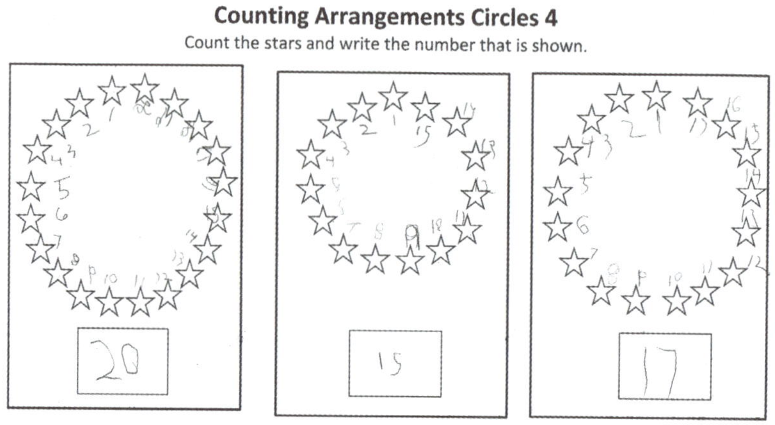

This student correctly counted the number of stars and wrote each number correctly. Looking closely, the student counted and wrote each number next to the stars. This strategy helped her keep track of the count.

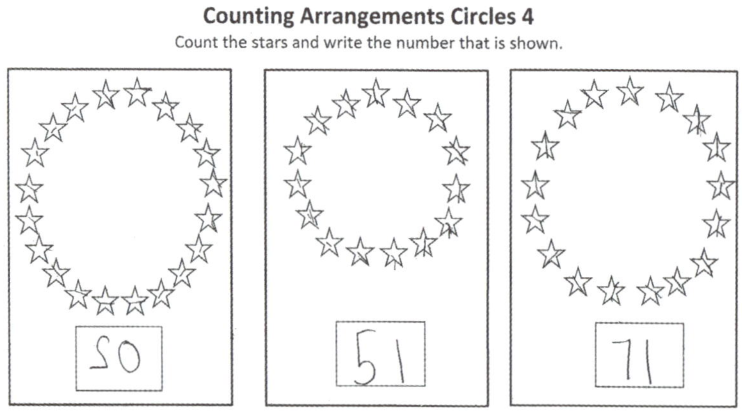

This student correctly counted the number of stars by marking each as she counted; however, she reversed the numbers as she wrote the total. It would be helpful for her to compare her written numbers with those on a number strip. Having a number strip nearby would allow her to see each number before writing it.

Additional Ideas for Support and Practice

The following ideas extend students' skills at counting groups of objects to 20 and provide meaningful practice with these skills.

Counting Page by Page

As numbers are introduced, students create pages to a counting book. The pages include:

- the number
- a double ten frame they fill in to show the number
- a statement to complete saying "__ is 10 and __ more"
- drawings showing the number in at least two different configurations.

The pages are saved and stapled together to form a book later in the year.

Counting Trains

Students count out cubes and build trains for numbers 11–20 and label the trains with number cards. Students identify that each number is 1 more than the previous number.

Prepare baskets of 155 connecting cubes and number cards for 11–20 for each group of four students.

Give each team a basket of connecting cubes.

Have students work in teams to select number cards and build trains for the numbers 11–20.

Work together to order the number cards in a row from 11–20.

Have students count as they touch each card in their row to check the order of their numbers.

Work with the other students on your team to build trains of cubes to show each number.

How many cubes would need to be in your train to show 13?

After you put the cubes together to form a train of cubes, place it on your table by the number card that matches.

Observe as students count out cubes, form trains, and lay them by the numbers in their row.

> Demonstrate how to lay the train on the table beginning at the number card.

When teams have made all of their trains and placed them in a row, ask:

What do you notice about the trains?

Are they the same size?

How is the size of each train changing?

Does that make sense?

Why do you think they are getting bigger for each number?
Do you see a pattern in the numbers?

Students might notice that the numbers have 1, 2, 3, and so on in them and go in order like the single-digit numbers in our counting sequence.

Have students write the numbers 11–20 across the bottom of a large sheet of paper and then create number towers to show them.

Number Beads

Students build strings of numbers 11–20 with beads on pipe cleaners to practice counting and to observe that each number is 1 more than the previous number.

Students put beads on a pipe cleaner (or string) to show the number 10.

Then they string 11 beads on another pipe cleaner and put them side by side.

Students continue to build numbers 12–20 to show that each number is 1 more than the previous number.

Students can draw a picture of the pipe cleaners and beads in their math journal and write the number to match.

Talk About It/Write About It

What is different about the numbers 1–9 and 11–19 when we write them?

How can you show 15 on double ten frames?

Sara says that there are more counters in the row than the circle. Is she correct? Explain.

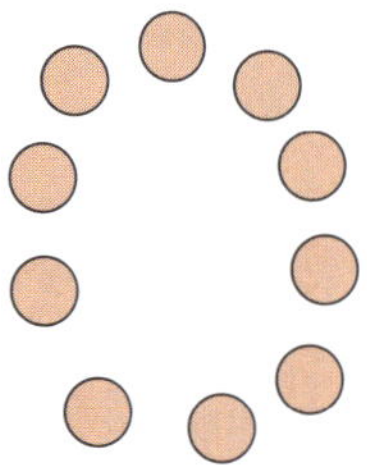

Liam had 10 and 4 more jelly beans. How many did he have? How do you know?

Bailey doesn't understand why 16 is more than 10. Can you tell her?

Are all teen numbers more than 10? Prove it.

Ordering Numbers

Students practice ordering numbers 11–20.

Show number cards 11–20 in random order on a document camera or interactive whiteboard.

We need to put these numbers in order.

What do you remember about numbers that could help you?

What number should we begin with?

Put *11* on the board.

Turn and talk to your partner about what number comes next. Be ready to tell me why.

Have students share their ideas and justify their choices for the next number as you continue to add numbers to build the 11–20 sequence.

Emphasize the familiar pattern in the ones digit that helps students figure out the order.

> **Differentiation**
> Some students may be ready to order numbers independently. Other students may benefit from using a number strip as a resource to order teen numbers.

Large Number Line

Students practice counting collections and then stand on the number to show the total.

Materials: large number line on the floor (1–20), bags of different amounts of manipulatives (11–20), blank double ten frames

Each student gets a bag of manipulatives and a double ten frame.

Students count the amount by placing the objects on the double ten frame.

Once completed, the student will find the amount on the large number line and stand on it. Once the teacher has checked for accuracy, the student may pick a new bag of objects to count.

Count and Color

Students color the number of items to match the number card they select.

Create templates with 20 objects on each that can be colored (e.g., 20 beach balls, 20 buttons). Objects should be in an array with 4 rows of 5 objects.

Place templates into sheet protectors and give students dry erase markers.

Students select a number card from a set of 11–20 and color that number of objects.

When they finishing coloring, they recount to check their work.

Then students erase, choose another number card, and color the new number of objects.

D Some More on Double Ten Frames

udents use counters to show teen numbers on double frames as 10 and ___ more and then draw pictures to represent the numbers.

Have students pick 4 number cards from a set of 11–19 cards.

For each number they choose, students show the number with counters on a double ten frame.

Then students draw the model on the 10 and Some More recording sheet and fill in ___ *is 10 and* ___ *more* for each number.

Bears at the Beach

In *Bears at the Beach* by Niki Yektai, the bears have bags, shells, pails, castles, and other items to count from 10 to 20.

Before Reading:

Give partners number cards 10–20 and have them place the numbers in counting order.

Count together as they touch each card to check the order.

We are going to read a story about bears at the beach.
Turn and tell your partner what you might see at the beach.

Have students share some ideas.

The bears have different numbers of each item.
As I read the story, listen to see if it mentions any of the beachy things you said and if the number order in the story matches the order of your number cards.

During Reading:

Read *Bears at the Beach: Counting 10 to 20* by Niki Yektai.

The book begins with 10 umbrellas and 11 bears and then continues to show groups of 12–20 beachy items. On every other page, a beach scene appears and students can find and count the items.

As you read, students should count the items on each page.

After Reading:

What kinds of beach items were in the story?
Did the number order match your cards?

Review that the teen numbers are a group of 10 and some more ones.
Read the story again having students say each teen number as 10 and ___ more ones.

Place a sheet protector over the pages so students can take turns circling a group of 10 and some more ones.

Place the book at a center for students to practice counting the items in arrays on the numbered pages and then find and count the items in the beach scenes.

Differentiation

The items in the beach scenes are in scattered configurations and although students do not need to count beyond 10 items in a scattered configuration, many students will love the challenge.

10 and Some More Links

SHOW IT!

Students use links to show numbers as 10 and some more ones.

Each student needs two sets of 10 links in two different colors (such as 10 red and 10 yellow) and a set of number cards 11–19.

Give students the links and place the set of number cards facedown in front of them.

Have students choose a number card.

Show how you can use the links to show 16 as 10 and some more.

Allow students time to work.

How did you show 16?

Show me the group of 10.

How many more ones do you have?

Choose another card to show another number.

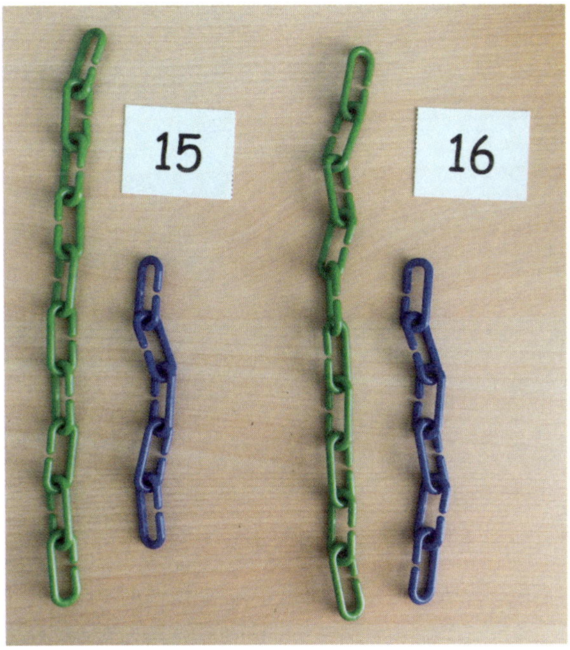

Students create linking chains to show the numbers on the cards. The links show a group of 10 and some more ones.

Making and Drawing Trains

Students show numbers as 10 and some more.

A student spins an 11–19 spinner and uses linking cubes to make a train of 10 and some more for the number.

In their math journal, students draw a picture to show the number and write:

My number is ___.

___ is 10 and ___ more.

If students are not able to copy the sentences, prepare templates to glue into their journals for them to fill in the blanks.

Eliminate It

Show the following graphic on the board.

Students work with a partner to decide which of the following does not belong with the others, then use math reasoning, numbers, or models to convince classmates that their thinking makes sense.

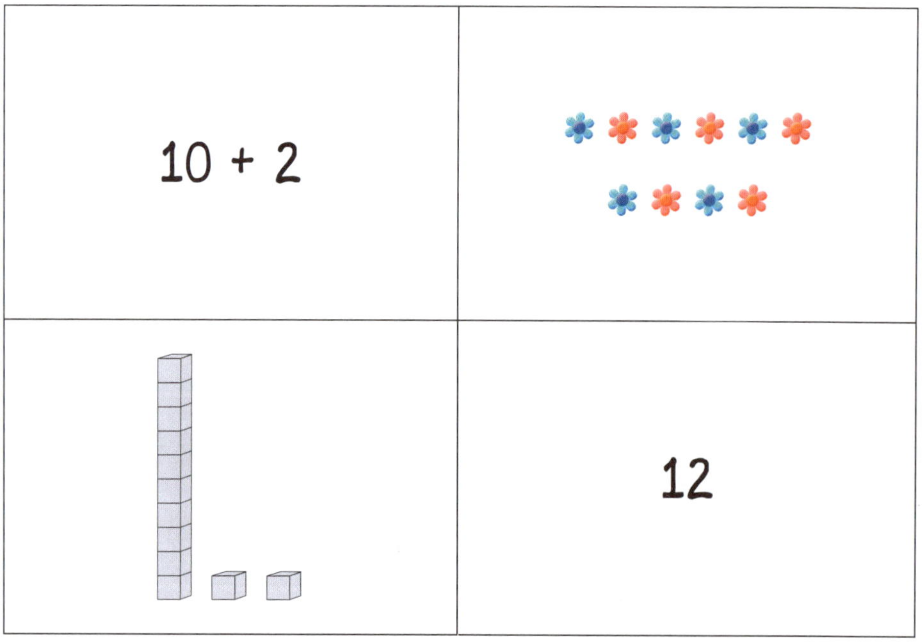

Double Ten Frame Match-Up

Students match double ten frame cards and number cards.

Materials: a set of double ten frame cards and 11–19 number cards for each pair of students.

Partners work to lay out the number cards in order, 11–19.

Then, players put the double ten frame cards in a stack facedown.

Player A selects a double ten frame card from the stack and matches it to the number card.

When Player A matches the number card, she says the number as a group of 10 and ___ more.

For example, if she matches the number 14, she will say, "14 is 10 and 4 more."

Then Player B selects a double ten frame card and matches it to the number card.

Partners take turns picking a double ten frame card, counting the dots to find the total, and then choosing the number card that matches it.

Players continue until all cards are matched.

Ten Frame/Number Concentration

This game is played like the traditional Concentration or Memory game. In this version, students take turns matching double ten frame cards with number cards for numerals 11–20.

Players need a set of small double ten frames and number cards for 11–20.

Place the double ten frames and number cards facedown.

Students take turns flipping over cards to find a match.

The player with the most sets of cards is the winner.

Differentiation

If this is too many cards for some students, modify the game to match fewer number sets like 11–16 or 16–20.

Counting Collection Partner Activity

Partners practice counting collections of various items and writing the number of each item.

Place collections of 11–20 objects in bags in the center area for students to count.

Player A selects a bag of manipulatives and counts them out loud.

Player B writes the number on a sticky note to match the set.

Players switch roles.

How Many?

Students practice counting 11–20 objects in different configurations by counting the number of stickers placed on index cards.

Create arrangements of 11–19 stickers on index cards (in rows, arrays, or a circle). Do not place in scattered configurations because the limit for scattered configurations is 10 objects.

Use a different type of sticker for each card (e.g., bunny stickers in a circle on an index card and duck stickers in a line on another card).

Students count the stickers and record their counts on a teacher-created recording sheet.

To allow you to check students' counting for accuracy, create recording sheets by placing one of each type of sticker and a line next to it where students record their count (e.g., they write the number of ducks by a duck on the recording sheet or write the number of bunny stickers by the bunny on the recording sheet).

Musical Math

Students show numbers 11–19 on a double ten frame as a group of 10 and some ones. They then write the number to match.

Materials: a bag of objects with 11–19 items, laminated blank double ten frames, dry erase marker, red yarn tied in a circle to fit around a ten frame for each student

Place sets of materials around the classroom.

Students walk around the carpet while music plays.

When the music stops, the students sit by the materials closest to them.

They place the objects on the double ten frame, filling one frame first.

Then they place the red yarn around any full ten frames.

They write what they did: __ *is 10 and* __ *more.*

For example, if the student is building 16, she would place the string around the top ten frame and count the items on the bottom ten frame, then write *16 is 10 and 6 more.*

When students are finished, they stand up to let the teacher know to check their work.

If correct, the student cleans up and waits for the music to play again.

Repeat as often as time allows.

Vocabulary

digit
double ten frame
leftover
eleven
twelve
thirteen
fourteen
fifteen
sixteen
seventeen
eighteen
nineteen
twenty

The spelling/writing of the number words is not critical. Instead, the focus is on knowing the number words that represent each quantity.

Make Writing Numbers Fun

Students need repeated practice writing numbers. Here are some ways to make writing numbers fun:

- gallon-sized bags with a small amount of paint spread out inside so students can trace numbers on the outside of the bags and see the form of the numbers
- $\frac{1}{4}$ inch of sand in the bottom of a plastic container or shoebox
- shaving cream on cookie trays
- make numbers with clay by rolling out snakes and bending them into numbers
- roll clay out flat and use the back of a plastic spoon to write on the dough.

ONLINE RESOURCES

General Resources

Additional Problems

Blank Double Ten Frame

Number Cards 11–19

Number Cards 11–20

Numbers 11–20 Student Observation Notes

Ten Frame Mats

Resources for Specific Activities

10 and Some More Recording Sheet

11–19 Spinner

Count More Stars Worksheet

Double Ten Frame Activity Sheet

Double Ten Frame Cards

How Many Dots? Student Sheet

Jelly Bean Ten Frame

Number Tents

These resources are available at http://hein.pub/MathinPractice, keycode MIPGK.

MODULE

4

Counting Numbers

About the Math K.CC.1; K.CC.2

One aspect of counting is knowledge of the counting sequence. Our students must be able to say the number names in order without skipping any numbers. The ability to count objects depends on our students knowing the counting sequence.

The key ideas focused on in this module include:

- counting to 100 by ones
- counting to 100 by tens
- counting beginning at any number.

In kindergarten, students learn this sequence, naming each number in succession without skipping numbers. While they memorize the sequence 1–20, they begin to see patterns that allow them to extend their counting beyond that without sole reliance on memory. Rather than memorizing the order of 100 numbers, students recognize patterns that make up the counting sequence, allowing them to continue with 21, 22, 23, 24, 25, 26, and so on (see Figure 4.1). These patterns help our kindergarten students master the counting sequence to 100.

Students in kindergarten practice counting by tens to 100. Because our number system is a base-ten system, this early skip counting practice gives students a foundational understanding that continues to deepen as they explore place value in first grade.

1	2	3	4	5	6	7	8	9	10
11	12	13	14	15	16	17	18	19	20
21	22	23	24	25	26	27	28	29	30

Figure 4.1 As students begin to notice the repetitive pattern in our counting sequence they are able to extend their counting beyond 20 by using this understanding.

Kindergarten students are also expected to count from any given number rather than always beginning at 1. Again, the patterns in our number system assist students with this skill. Hundred charts make these patterns visible as students explore the repetition of the counting sequence. Students recognize that if they begin with 35, the number sequence continues with 36, 37, 38, 39, following the 0–9 pattern through the thirties.

A particularly problematic component of the counting sequence for some students is counting across decades from 19–20, 29–30, 39–40, 49–50, and so forth. Exploring patterns by looking at a hundred chart can help to spur insights about next numbers (see Figure 4.2).

1	2	3	4	5	6	7	8	9	10
11	12	13	14	15	16	17	18	19	20
21	22	23	24	25	26	27	28	29	30
31	32	33	34	35	36	37	38	39	40
41	42	43	44	45	46	47	48	49	50
51	52	53	54	55	56	57	58	59	60
61	62	63	64	65	66	67	68	69	70
71	72	73	74	75	76	77	78	79	80
81	82	83	84	85	86	87	88	89	90
91	92	93	94	95	96	97	98	99	100

Figure 4.2 As students count along following the hundred chart, they notice that twenties are followed by thirties and thirties are followed by forties.

Learning Goals

I can count to 100 by ones.

I can count to 100 by tens.

I can begin counting at any number.

Lessons in This Module

Rote Counting by Ones
Build a Class Hundred Chart
Count Every Day
The 100th Day of School

Counting by Tens
Clothesline Number Tents
Zero

Counting on from Any Given Number
What Comes Next?
Crossing the Decades

Counting is more than knowing the counting sequence. It is also about understanding numbers and concepts about numbers (e.g., one-to-one correspondence and cardinality). There are many connections between Modules 1–4 because these modules all explore important ideas about counting. As kindergarten students refine their knowledge of the counting sequence, they will also be counting objects (Modules 1–3) and gaining insights about counting numbers, such as each consecutive number in the counting sequence represents 1 more than the number just before it. Activities to master the counting sequence are stretched across the kindergarten year with students counting to 20 early in the year and then gradually increasing their counting to 50 or 80, and ultimately 100. Gradually we introduce students to counting by tens and counting from any given number. And although activities related to the counting sequence appear in this module, students simultaneously explore related counting standards in other modules (e.g., counting objects and reading and writing numbers 1–20).

Exploring the Progression

PREVIOUS		NOW		NEXT
Preschool		**Grade K**		**Grade 1**
Rote counting to 20, with minimal attention to number patterns; counting usually begins with 1	»	Counting to 100 by ones and tens; counting beginning at any number; exploring number patterns	»	Counting to 120; exploring number patterns beyond 100; transitioning from counting all to the addition skill of counting on (6 + 2: begin at 6 and count 7, 8)

Common Error

Students may have trouble with the counting sequence of the teen numbers; 11, 12, 13, and 15 do not follow the language pattern of 14, 16, 17, 18, and 19. Students may also have trouble remembering the words twenty, thirty, and fifty because they do not follow the language pattern of the decades like forty, sixty, seventy, eighty, and ninety.

Ideas for Instruction and Assessment

ROTE COUNTING BY ONES

Rote counting is a foundation for kindergarten mathematics. Students learn to count by counting every day. Following are a variety of ways to keep your students counting by ones.

A completed hundred chart may be posted in another place in the classroom, but this one is built with students and used to explore and predict the number sequence.

Build a Class Hundred Chart SHOW IT!

Building a hundred chart, adding a square for each day beginning with the first day of school, provides a visual that is used throughout the year for daily counting practice and the exploration of numbers and patterns.

Each day, post the day's number on the wall to create a class-made hundred chart.

Have students count each day from 1 to that day's number, pointing to each number as they count.

Differentiation

In first grade, students will rote count to 120. Some kindergarten students may be ready to extend the counting sequence beyond 100 to 120.

After day 10, post the 11 in a new row, under the 1.

As the days pass, discuss the patterns that are appearing.

What do you notice about the numbers?

Can you predict what number tomorrow will be?

Color-coding can make patterns more visible (e.g., the numbers in the column with 1, 11, 21, 31, 41, 51, and so on are all written on blue squares and the numbers in the column with 2, 12, 22, 32, 42, and so on are written on yellow squares).

Count Every Day

Have students practice with the counting sequence as often as possible.

Count to start the day, count as you are in line to go to lunch, count as you get jackets at the end of the day.

Make counting fun. Try counting in these ways:

- in whispers
- in funny voices
- while you hop
- as you add counters to a jar
- have a student pick a number from the hundred chart and count to it.

If students begin to count very quickly, ask them to slow their pace and attend to precision. Students often make mistakes if they attempt to count too quickly.

FORMATIVE ASSESSMENT

Ask students to count orally from 1–100.

Mark the first number that is incorrectly stated.

Record observational notes of student responses as he or she counts.

If a student is not able to count past 20, revisit counting by using a number strip or part of a hundred chart to 20.

Introduce each decade and revisit the counting pattern for each.

When the student is able to count to 30, begin moving on with each decade to 50 and then 100.

Use the observational notes page.

Differentiation

Students will move through the counting sequence at different rates. Some may immediately see patterns and transfer the patterns to the whole counting sequence, and others may take longer to do that. Continue providing visuals (hundred charts, number lines) and talk about the patterns. Although daily whole-class counting activities are important, it is beneficial to work in small groups with students who are having difficulty with the counting sequence. Understanding the 1–20 sequence is foundational.

It is important to introduce students to sets of numbers as they learn to rote count. Begin counting to 10, then to 20 and 30 as students learn the patterns in the counting sequence. Then count to 50, 70, and 100 to extend the pattern.

Some of these ideas mix rote counting with one-to-one correspondence (e.g., counting as you put counters into a jar). Knowing the counting sequence and the ability to count objects are both important skills that need attention, but they also are skills that are integrally connected.

For more ideas on integrating children's literature into math lessons, see Math in Practice: A Guide for Teachers, Chapter 2.

The 100th Day of School

Students celebrate the 100th day of school by listening to a story, *The 100th Day of School* by Angela Shelf Medearis, and counting to 100.

Materials: collections of 100 items for students to observe (pennies, pom-poms, books)

Before Reading:

Briefly set a context for reading the book.

We have been counting the number of days of school since the very first day.

Where do we keep track of the number of days of school?

What number was yesterday?

So what number is today?

Today is the one-hundredth day of school!

Add the number 100 to the class hundred chart.

During Reading:

Begin reading *The 100th Day of School.*

Pause as you read to discuss how the students in the story also keep track of the number of days of school and count to 100.

Point out the different groups of 100 items that the students collect.

After Reading:

The students in the class counted to 100. Let's count to 100!

Stand near the class 100 chart and touch each number as the class counts.

Show the students a clear bag of 100 pennies.

How many pennies do you think are in this bag?

This is what the 100 pennies look like.

Students in kindergarten are not expected to count 100 items. By the end of the year they should be able to rote count to 100 by ones and tens.

Pass the bag of pennies around for students to examine.

Look at the 100 pennies and feel how heavy they are.

Show the students a clear bag of 100 pom-poms.

How many pom-poms do you think are in this bag?

This is what 100 pom-poms look like.

Pass the bag of pom-poms around for students to examine.

Look at the 100 pom-poms.

How do they feel compared to the 100 pennies?

Note the connections to measurement. Students are using attribute words like heavier or lighter in context.

Then show the students 100 books that are stacked.

How many books do you think are stacked in front of the room?

This is what 100 books looks like.

We can't pass around 100 books. They would be hard to hold.

Sometimes 100 is small like 100 pennies or pom-poms and sometimes 100 is big like a stack of books.

But it is still 100.
What would 100 elephants be—big or small?
What would 100 ladybugs be—big or small?

Have students turn to partners and talk about things that would be big if you had 100 and others that would be small if you had 100.

Write a list of both on the board.

Students then write or draw in their math journals and complete the sentences:

100 _______ are big.
100 _______ are small.

Additional Ideas for Support and Practice

The following ideas extend students' skills at counting by ones and provide meaningful practice.

Partner Count

Students alternate counting with a partner.

Students work with a partner to practice counting to a given number.

Player A starts at 1. Player B says 2. Player A says 3, and so forth.

Partners continue taking turns counting to 10, 20, 50, or 100.

As students progress with counting, they can begin counting from numbers other than 1.

Count Around the Circle

Students sit in a circle to count.

Have students sit in a circle to practice counting.

All students count quietly in a whisper voice, but at each child's turn, he says the number in a louder voice.

As the class gains experience with counting, you can begin counting at any given number.

A variation of this game is Tens Pop. Students count around the circle. When a student gets to 10, they say the number and pop up. Then the counting continues, popping up at each 10 until the group has counted to 100.

How Many Hands?

Students count to find the total number of hands for 5, 10, or 20 students.

How many hands does 1 student have?
How could we find out how many hands 5 students have?

Have 5 students stand up and count the number of hands.

Continue counting hands with 10 students and then 20 students.

How many hands are in our class?

Variations include counting by ones for the number of smiles or noses.

Students can count by tens to count the number of fingers or toes.

Clay and Sticks

Students order numbers written on craft sticks in clay.

Materials: clay and craft sticks with numbers 1–10

Write the numbers 1–10 near the end of a craft stick, one number per stick.

Students make a clay snake and then arrange the number sticks from 1–10, sticking each in the clay in the correct order.

This student works to order numbers 1–10 by placing the sticks in clay. This activity could be adapted to count to 20 or to count by tens by labeling the sticks 10–100.

Count the Number of Doors

Students could count chairs, bulletin boards, or other items within the school.

Students count doors as they walk through the hallways.

Take a walk through the school counting doors along the way.

You might count the doors on the way to lunch or music or PE.

COUNTING BY TENS

Students practice counting by tens and explore the patterns they see when counting by tens.

Clothesline Number Tents SHOW IT!

For a small-group lesson, hang a clothesline or piece of yarn at the students' level. Fold pieces of paper in half to create number tents and write a ten's number on each tent (10–100). Students place the numbers in order on the string.

Have a hundred chart nearby.

Sometimes when we count, we count by tens.

What do you remember about tens? (They are at the end of each row on the hundred chart. *Or,* They are a full tens frame.)

Do you know any numbers that count by tens? (10, 20, 30, . . .)

Place the number tents in front of the students in random order.

What do you notice about these number tents? (They are all tens.)

Check to be sure that the students can name each number.

How can we place these number tents in order on our number line?

What tens number should we start with?

Have a student find 10 and place it on the clothesline.

What number would come next?

How do you know?

Have a student find 20 and place it on the clothesline to the right of 10.

This activity can also work for the counting sequence, number patterns, and crossing decades.

Differentiation

For additional support, some students may just need to work with numbers 10–50 and have access to a hundred chart to use as a reference.

Number tents can be ordered on yarn hung below the board ledge. Students can practice the counting order, counting by tens.

Have students read the numbers as you point to 10, 20, and so on.

Continue to have students place numbers to 100 on the clothesline.

Each time a new number is placed, have students read the tens that are on the line.

> *Turn and tell someone next to you something you noticed about all of our tens numbers.*

Give students number cards 10–100 and have them work with partners to order and count them.

FORMATIVE ASSESSMENT

Tell the student to count by tens to 100.

Listen as the student counts.

Record observational notes as the student counts.

Use the observational notes page.

If the student is not able to count by tens, revisit the counting pattern of the tens and have her continue to practice.

While practicing, it is helpful to use a chart or number cards. The student should touch and say each number when she counts. This provides movement (as she touches) and visual (as she reads) and auditory (as she says each number) reinforcement.

Frequent practice in counting by tens helps students to learn the pattern. This girl uses a pointer to touch each number to count by tens. With a hundred chart she can also practice counting on from a given number or crossing a decade.

Zero

In the book *Zero* by Kathryn Otoshi, students explore how 0 has no value until it is matched with another number as in 10, 20, 30, and so on.

Before Reading:

Make sure everyone is familiar with the concept of 0.

What is 0?

If I have 0 pieces of candy, how many do I have?

What is another way to say 0?

During Reading:

As you read, stop periodically to discuss Zero's problem in the story (i.e., he is upset because he does not have any value).

After Reading:

Give each student a pack of 0–9 number cards that are mixed and not in counting order.

There are 10 cards in your set.

Show how you would put these cards in counting order.

Observe as students are positioning their cards.

Once all students have ordered their cards, ask for volunteers to explain how they ordered their cards and where they placed 0 in the order.

How did you place your cards in order?

Why did you place 0 at the beginning of the sequence?

If you showed me the value of 0 with counters, how many counters would it be?

Record on the board: *0, 1, 2, 3, 4, 5, 6, 7, 8, 9.*

Use your cards to make the number 10.

How did you show 10?

How many counters would show the value of 10?

So 10 has value even though it has a 0 in it?

Record *10* on the board just below the 1.

Can you make 20 using your cards?

How did you show 20?

How many counters would show the value of 20?

So 20 has value even though it has a 0? If I had 20 apples, would I have apples?

Record *20* next to the 10 on the board.

Continue asking students to use their cards to make 30–90 and continue recording the numbers on the board just below the 0–9 numbers.

0, 1, 2, 3, 4, 5, 6, 7, 8, 9

10, 20, 30, 40, 50, 60, 70, 80, 90

What number would come next?

Have students count with you as you count by tens from 10–100.

What do you notice about the numbers we created?

Have students write the numbers 10–100 in their math journals.

Additional Ideas for Support and Practice

The following ideas extend students' skills at counting by tens and noticing the patterns and provide meaningful practice.

Highlighted Tens

A hundred chart is filled with numbers. Highlighting the tens on a hundred chart helps students notice these numbers.

On the class hundred chart, highlight the tens with a different color of ink or by overlaying pieces of acetate squares (colored transparent plastic squares).

Then have students count by tens as they focus on the highlighted squares.

Count!

In *Count!* by Denise Fleming, colorful illustrations count by ones from 1–10 and then count by tens from 10–50.

Before Reading:

Have the students review counting by ones to 10.

During Reading:

Read *Count!* As you read and count, ask the students:

What number will be next?
How do you know?

After Reading:

Say to the students:

The author stopped counting by tens at 50.
What number comes after 50 when we count by tens?
Let's count to 100 by tens.

Counting Ten Frames

Students count by tens with ten frame cards.

Option 1

Students work with partners and take turns counting to 100 by tens.

Students use 10 ten frame cards (each representing 10), setting 1 down each time they say a number as they count.

Partners give a thumbs-up if the counting sequence is correct.

Option 2

For a partner game, make cards labeled 10–100 and place them in a paper bag.

Give each set of partners a paper bag of number cards and a set of 10 ten frame cards (each representing 10).

Partner A closes his eyes and pulls out a number card from the bag, then uses the ten frame cards to show the number he chose.

Player B checks his work by saying the number that was chosen (e.g., 60) and counting the 6 ten frames.

Then player B picks a card.

Column Hundred Chart

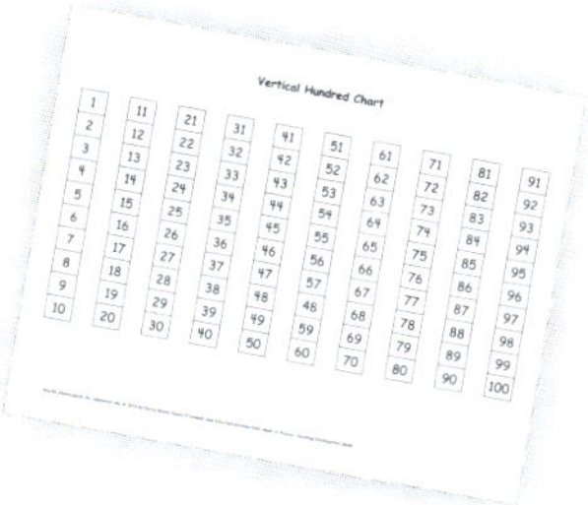

An alternative to the traditional hundred chart is a vertical, or column, hundred chart that shows the decades in columns. A vertical hundred chart can help students see the counting patterns in a different way.

The pattern of the decade is visible on the left of each column and the counting sequence continues down the column. Students can count to 100 by tens by reading across the bottom.

Count the Frames

Students count collections of ten frames (each representing 10).

Set up 10 stations around the room, each having a collection of 1–10 ten frame cards (i.e., one station has one card, another has two cards, another has three cards, and so on). Each student needs a hundred chart to record the number at each station.

Students move with partners to count the numbers (by tens) and say the number they counted. Before moving to the next station, students circle that number on a hundred chart.

By the time they have visited all stations, they should have circled all 10 of the tens numbers on their hundred chart.

Painting 100 Dots

Students use cotton swabs to paint 100 dots.

Divide a paper into 10 sections.

Students use different colors of paint and cotton swabs to paint 10 dots in each section for a total of 100 dots.

Painting can be done over 10 days, with a different color and section for each day. Students can then touch each section as the class practices counting to 100 by tens.

Cover the Tens

Students use transparent chips to cover tens to explore patterns when counting by tens.

Have students use transparent counters to find and cover tens on a hundred chart.

They practice counting by reading the column of tens.

COUNTING ON FROM ANY GIVEN NUMBER

Students practice counting to 100 starting at any number. This skill prepares them for counting on and identifying 1 more without counting all.

What Comes Next?

Students explore patterns in numbers to figure out what number comes next in the counting sequence.

In small groups, write and say three numbers in the counting sequence (e.g., 22, 23, 24) and ask students

What comes next?
How do you know?
Is it still a 20? Why?

Have them find the number on the hundred chart to check.

Repeat for several number sequences.

Begin posing just one number and ask what comes next.

Ask students to explain their thinking.

Find the number on the hundred chart to check their ideas.

Thinking Through a Lesson

Crossing the Decades

In a small group, students explore patterns to find the numbers that follow 19, 29, 39, 49, 59, 69, 79, 89, and 99 in the counting sequence.

Write *29* on the board.

If I started counting at 29, what number would I say next? (30.)

Crossing decades from 29–30 or 59–60 is a very difficult part of the counting sequence for many students. Understanding number patterns allows students to make sense of what comes next.

Have students turn and tell their partners what comes next and why. Listen to their ideas.

SMP3

Would the next number be a twenties number? Why or why not? (No because there aren't any more twenties numbers.)

What do you mean? (29 is the last one because it has a 9, you can't say *twenty ten*; that's not a number.)

So what would you say? Tell your partner. (30.)

How do you know that 30 comes next? (Because after twenties is thirties and you are done with twenties after 29.)

Recognizing that any number ending with a 9 is the last number in that decade is an important realization.

Have a student check the hundred chart to confirm that 30 comes after 29.

Focusing on the patterns in the tens digits allows students to figure out which number comes next. After twenties is thirties and after thirties is forties. . . .

Write 49 on the board.

If I started counting at 49, what number would I say next?

Continue the same line of questioning having students justify how they know there are no more forties and why 50 comes next. Have someone confirm it on the hundred chart.

Relating the order back to the hundreds chart shows students how they can check their own work when deciding on the next number and allows students to see the patterns on the chart.

SMP5

When you are counting and a number ends in 9, what should you think about? (We can look at the tens on the hundred chart to see which 10 is next.)

What if you didn't have a hundred chart to look at? How would you know which 10 is next? (We could think about counting by tens: 10, 20, 30, 40, 50, and so on.)

Highlighting the nines and tens numbers on the hundred chart and observing the patterns helps students see this repetition.

SMP7

After a few more examples, show students the following number cards:

9, 19, 29, 39, 49, 59, 69, 79, 89, 99

10, 20, 30, 40, 50, 60, 70, 80, 90, 100

Connect this to counting by tens. Students may need to count aloud by tens for a bit until the pattern is evident to them.

Shuffle the nines cards and lay them facedown on the table; lay the tens cards faceup.

Go around the group and have each student pick a nines card and say what comes next and why.

This allows you to assess which students have internalized the ideas of how to find the next number and which need additional support.

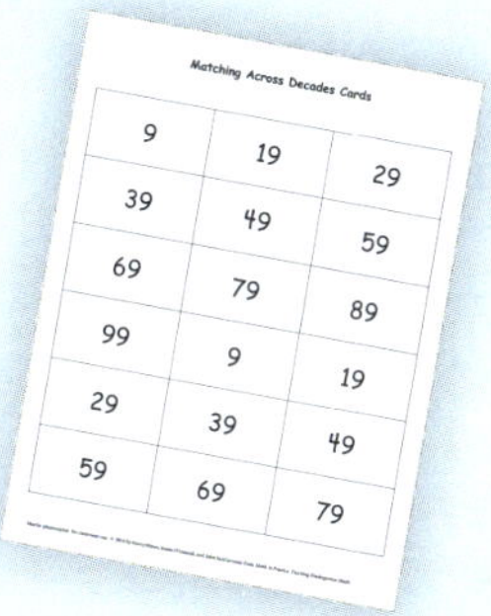

Matching Across Decades Cards

9	19	29
39	49	59
69	79	89
99	9	19
29	39	49
59	69	79

SMP3

Having other students "check" the next number keeps the group actively involved in the lesson.

Students then match the nines card with the tens card that comes next.

Other students in the group give a thumbs up if they agree with the number the student says comes next.

Rather than correcting a student who chooses the wrong number, let other students think about it and tell their ideas of what should come next.

If they disagree, they show a thumbs down and explain their thinking.

Have students refer to the hundred chart to check their answers.

Place the cards at the center area so students can practice matching the nines cards and the tens cards at their next visit to the center.

Placing the cards at the center area allows students to do the matching game with partners and get additional practice matching the numbers that cross decades.

Observe where students experience trouble to be able to offer practice that fits their needs.

Are they having difficulty with numbers in the middle of a sequence (e.g., 24–25)? Focus on the patterns in a decade (e.g., how the numbers progress from 20–29, emphasizing the 1–9 progression).

Are they having difficulty crossing the decades (e.g., 19–20)? Focus on the patterns across a decade (e.g., What happens when you are done with twenties? What decade comes next?).

Provide students with hundred charts and have them highlight and discuss the pattern that relates to their error.

FORMATIVE ASSESSMENT

Have students begin counting at different numbers, allowing them to cross a decade.

Start counting at 24. Stop the student at 41.

Start counting at 50. Stop the student at 75.

Start counting at 82. Stop the student at 100.

Using these numbers allows students to count on from a given number and cross decades.

Record observational notes of student counting.

Use the observational notes page.

Additional Ideas for Support and Practice

The following ideas extend students' skills at counting on from any given number and provide meaningful practice.

Count from a Number on the Hundred Chart

Practice counting on from a given number by randomly selecting a number on the hundred chart.

> Stand in front of the class hundred chart with your eyes closed.
>
> Randomly point to a number and the class starts counting on from that number.
>
> As the class becomes familiar with the routine, students can point to the number.

Missing Numbers

> Provide students with five frames that have some numbers filled in and others that are missing.
>
> Have them fill in the frames to show the correct counting sequence.

Common Error

Many children reverse individual numerals and/or the digits in a two-digit number as they are learning to write them. Post the numbers in a visible place. Ask, "Does your number look like the number on the chart?"

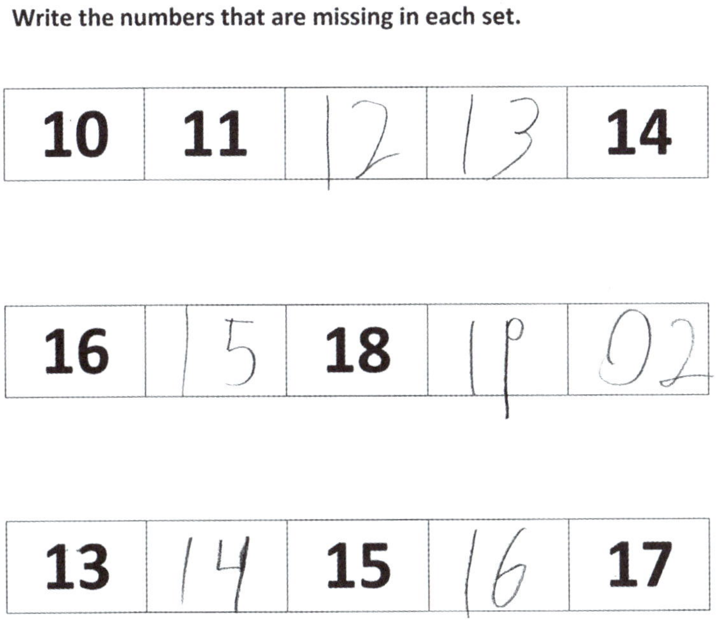

Write the numbers that are missing in each set.

10	11	12	13	14

16	15	18	19	02

13	14	15	16	17

This student shows her ability to count and fill in the missing numbers in the first and third sets of numbers. She makes an error in the second set of numbers, placing 15 after 16. Having her read her number sequence out loud might allow her to hear her own mistake. Having her check her sequence with the order on a number strip or hundred chart allows her to find her error. Have her explain the error to you. Note: She counted 19 and 20 correctly, but reversed the order of the digits when she wrote them. Although this is quite common at this level, remind her of the reversals.

Counting On from a Number in a Circle

This is a variation of Count Around the Circle, with students sitting in a circle to practice counting on from a given number.

Students sit in a circle and start counting from a given number:

We have 24 students in our class. Let's start counting at 24.

Today is the thirty-first. Let's start counting at 31.

Emma and Rob ride bus number 58. Let's starting counting at 58.

We have been in school for 63 days. Let's start counting at 63.

All students count quietly in a whisper voice but at each child's turn, she says the number in a loud voice.

Talk About It/Write About It

Jamie was counting and got to 34. Then she forgot what comes next. Can you help her? Explain what comes next and how you know.

What number comes after 29 when you count? How do you know?

Move and Count

Students count to 100 and do 100 jumps or claps. The class can do 100 jumps, claps, stomps, jumping jacks, and so on.

In the beginning, you might stop at 10 or 20, but as students gain more counting experience, stop at 17 or 32 so students have to begin the next counting sequence at those numbers.

Begin counting at 1 and do 10 claps.

Take a break but throughout the day, do some more claps beginning where you left off, keeping track on the board of how many you have completed.

Make it a goal to get to 100 by the end of the day!

Popcorn Counting

Teams of students practice counting on from a given number.

Split the class into four or five teams.

A team starts the counting and continues to count until the teacher gives a signal (e.g., a bell, a clap).

The teacher then points to another team that picks up the counting where the last team left off.

Teams pay attention so they know when it is their turn to do the counting.

Partner Picks

Students take turns picking a number on a hundred chart and asking their partner what comes next.

Partner A has a small hundred chart and picks a number.

Partner B has to count on five more numbers while Partner A checks it on the hundred chart.

Partners switch roles and do it again.

Interactive Hundred Charts

A hundred chart with pockets can provide many ways for students to practice counting by ones or tens, crossing decades, and identifying patterns. Interactive whiteboards also allow students to manipulate numbers on a hundred chart.

Ideas here span all of the topics related to rote counting.

Use a hundred chart with pockets, or other interactive hundred chart, as a warm-up with the class or set up as a center for independent practice:

- Students can use a pointer to touch each number as they count.
- Remove random number cards and have the students replace them.
- Place acetate squares over the decades and have students practice counting by tens.
- Remove numbers that cross decades (e.g., 29 and 30, 39 and 40) for students to replace.
- Remove a partial row of numbers (52–60) for students to replace, using the counting pattern or counting on from a given number.

Differentiation

Some students benefit from counting with a pause as they take a second to think about the next number: 27, 28, 29 . . . 30.

Flash Ten at Decades

As you practice counting each day, make students aware of each decade by flashing 10 fingers as you say 10, 20, 30, and so on.

Matching Across the Decades

Students match numbers that cross decades.

Copy and cut the Matching Across the Decade cards 9, 19, 29, 39, 49, 59, 69, 79, 89, 99 on one color of paper.

Copy and cut the cards 10, 20, 30, 40, 50, 60, 70, 80, 90, 100 on another color of paper.

Students spread out the tens cards and then shuffle the nines cards placing them in a pile, facedown.

Students work with partners to pick a nines card and match it with the tens card that comes next.

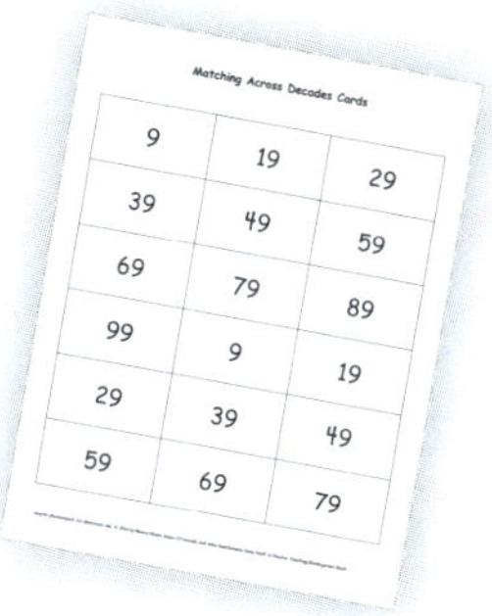

Cross the Decade Partners

Students use numbers written on large pieces of paper to find their cross-the-decade partners.

Write each number—9, 19, 29, 39, 49, 59, 69, 79, 89, and 99—on a full-size piece of paper of one color.

Write each number—10, 20, 30, 40, 50, 60, 70, 80, 90, and 100—on a full-size piece of paper of another color.

Give each student a number.

Give each student a card that matches another student. If there are more cards than students, keep the extra cards separate. At the end of the activity, you can show those cards as you check the numbers on the hundred chart.

Students find and stand with their partner, who is holding the number before or after (e.g., 9 finds 10).

Play music as the students find their partners.

When all number pairs have been found, have partners say their numbers and check them using the hundred chart.

Hopping Number Count

Students hop on number squares on the sidewalk to count and cross decades.

It would be helpful to draw several sets of squares so all students can participate.

On the playground, draw 20 squares that are connected in a 2-by-10 array and are large enough for a student to stand in.

In the first set of two squares, write *9* and *10*.

Continue writing a pair of numbers to cross the decades in each set of squares (19 and 20, 29 and 30, and so on).

99	100
89	90
79	80
69	70
59	60
49	50
39	40
29	30
19	20
9	10
student starts here	

Vocabulary

count
hundred chart
ones
order
pattern
sequence
tens

Version 1

A student starts in the square with 9 and says, "9."

He jumps to the 10 square and says, "10."

Then he jumps to the 19 square and says that number.

The student continues to jump in each box and say the number to practice crossing the decade.

Version 2

A student starts outside of the square with the number 9 and counts from 1–8.

When she says 9, she jumps in the square with the 9 and then jumps to the 10 square as she counts the number 10.

The student continues to count from 11–18 while standing in the 10 square.

When the student counts 19, she jumps to the 19 square and then jumps to the 20 square as she counts the number 20.

The student continues to count the decade on the tens square and jump squares to cross the decade.

ONLINE RESOURCES

General Resources

Additional Problems

Counting to 100 Student Observation Notes

Hundred Chart (horizontal and vertical formats)

Number Cards 0–9

Number Cards 10–100

Ten Frame Cards

Resources for Specific Activities

Matching Across Decades Cards

Write the Missing Numbers 10–20

These resources are available at http://hein.pub/MathinPractice, keycode MIPGK.

MODULE

5

Comparing Numbers 1–10

About the Math K.CC.6; K.CC.7

Students are beginning to understand that numbers represent a quantity. They have counted groups of objects and can name the quantity in a set (e.g., 5 stickers or 8 erasers). As students explore numbers, they create models of the numbers. They have made towers of connecting cubes or placed counters on ten frames to visualize quantities. They are noticing, with our help, that when they count, each successive number they say represents a quantity that is 1 more than the previous number said. Now, students use this information as they begin to compare quantities and numbers.

The key ideas focused on in this module include:

- comparing two quantities to decide which is more or less by using manipulatives or drawing pictures
- comparing two quantities to decide which is more or less by looking at the numbers.

When comparing groups of objects, students decide which group has more or fewer objects than the other. Counting and matching are two of the strategies students use. As students compare the heights of towers of cubes by placing them next to each other, compare rows of counters by matching them side by side, or count to compare the number of spaces filled by counters on a ten frame, they are making decisions about which number is greater or less (see Figure 5.1).

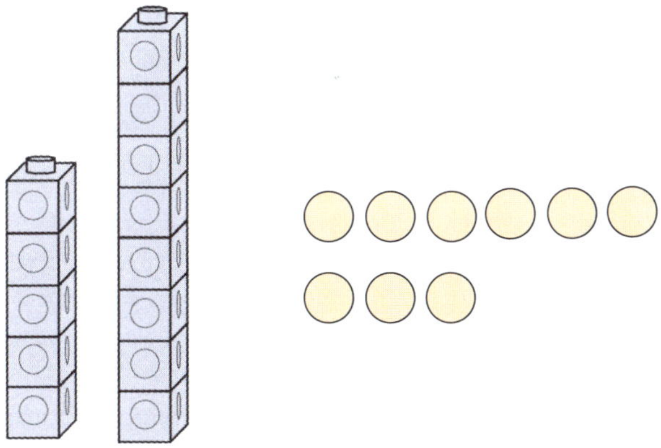

Figure 5.1 Students line up objects to compare quantities.

Students use matching and counting strategies to compare groups of objects, but after those experiences with objects, they are challenged to determine more or less when comparing numbers 1–10 without the objects.

Understanding the counting sequence, and the very important concept that each successive number is 1 more than the number before it, allows our students to make decisions based solely on the numbers. Looking at the order of numbers on number lines (see Figure 5.2) and hundred charts supports students as they internalize this concept.

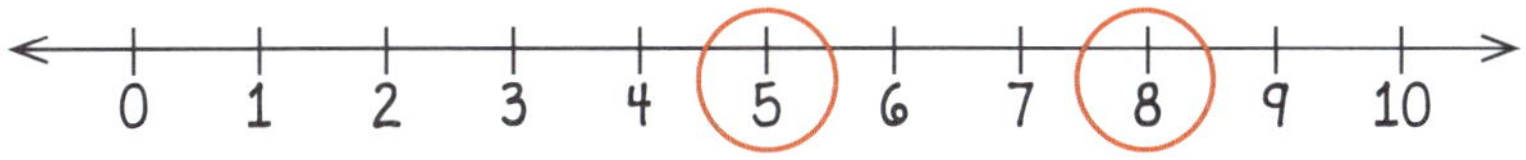

Figure 5.2 Students circle numbers on a number line to compare them.

Exploring the Progression

PREVIOUS		NOW		NEXT
Preschool		**Grade K**		**Grade 1**
Possible familiarity with the words *equal to, more* or *less*, though unable to express the meaning	»	Comparing two quantities, first by using manipulatives or drawing a picture and then by looking at the numbers	»	Comparing two-digit numbers; learning the symbols for greater than (>) and less than (<)

The timing for this will vary depending on the student, but the goal is for students to be able to compare numbers 1–10 by the end of the kindergarten year.

Learning Goals

I can tell which set has more or fewer objects.

I can tell which number is greater or less.

Lessons in This Module

Comparing Groups of Objects

Comparing in the Classroom

Towers of Cubes

Matching Bears

Just Enough Carrots

Using a Balance to Compare

Building Numbers on a Number Strip

Comparing Written Numerals 1–10

Number Line Circles

Target Number

Ideas for Instruction and Assessment

COMPARING GROUPS OF OBJECTS

Students' initial experiences with comparing quantities help them make sense of the terms *more*, *less*, and *the same*. Students compare groups of objects to determine which group has more or less and draw pictures to compare quantities.

Comparing in the Classroom

It is grammatically incorrect to say there are less objects in a set. There are actually fewer objects in a set, but we say that one number is less than another number. With that said, young students often understand the idea of more or less and so using both less and fewer is acceptable at this level, as long as we make sure to make the connections between fewer/less than and more/greater than.

This activity uses everyday classroom objects to help develop the meaning of the terms *more*, *less/fewer*, and *same*.

Show students 6 blue cubes and 4 red cubes.

Turn and ask your partner: Are there the same number of blue cubes as red cubes?

Is there more of one color? Which color?

What does it mean to have more?

How do you know there are more blue cubes?

Would it help if I line them up next to each other? Why?

Show students 3 pencils and 5 crayons.

Are there more pencils or crayons?

What could you do to help you decide which group has more?

Consistently ask students questions to clarify their thinking. If a student responds that pencils has more, ask them why rather than quickly dismissing the answer. It may be that he was thinking about the length of all of the pencils combined. Through continued questioning, you may see the answer in a new light, and your questioning prompts students to be precise in their talk about math ideas.

Have 2 girls and 3 boys stand up.

Do we have more girls or more boys standing?

What could you do to help you decide which is more?

Record the word *more* on the word wall or Math Talk chart. Draw a row of 5 hearts and a row of 3 hearts. Ask students which is more and then circle the 5 hearts to show more.

Continue with examples of *more*.

If students are easily identifying *more*, continue with the following *less/fewer* and *same* examples. If not, continue working on *more* and address *less/fewer* and *same* at another time.

Differentiation

Many students arrive in kindergarten understanding the concepts of more, less, and same, but others may need repeated examples to develop the concept and language.

Show students 6 blue connecting cubes and 3 red connecting cubes.

Which group has less?

What could you do to help you decide which group has less?

What does it mean to have less?

We could also say we have fewer red cubes.

Record *less* and *fewer* on the word wall or Math Talk chart. Draw a row of 5 stars and a row of 3 stars. Ask students which is less/fewer and then circle the 3 stars to show fewer.

Show students two cups, each holding 3 pencils.

Which cup has more pencils?

Hold up one of the cups.

So this cup doesn't have more? And it doesn't have less?

What do we call that?

What does it mean if we say we have the same number of pencils?

Record the word *same* on your word wall or Math Talk chart. Draw a row of 4 happy faces and another row of 4 happy faces. Ask students which row has more and then circle both rows to show the same.

The formal terms of greater than, less than, and equal to are introduced in a later lesson.

Continue with examples as needed.

To conclude the lesson, provide a group of objects that all students can see (in the middle of a group seated on the floor, using a visualizer, among other ways).

Ask volunteers to divide the objects into two groups.

Which group is more? Explain.

Which group is less? Explain.

Are the groups the same? Explain.

Differentiation

For students who quickly understand more, less, and equal to, ask them to create their own groups of items to demonstrate each vocabulary term.

This math talk will provide students with more opportunities to gain understanding from other students' thinking.

Towers of Cubes SHOW IT!

Students build towers with more or less than a specified number of cubes.

Create a tower with 6 cubes. Have students count the cubes in the tower you make.

Can you build a tower that has the same number of cubes as mine?

Observe students as they build.

How do you know yours is the same as mine?

Can you make another tower, this time with more cubes?

Observe students as they build.

How do you know this tower has more cubes than your other tower?

Ask students to count the number of cubes they have in their new tower and share the number.

Record the numbers on the board. Talk about each one of them as you record it.

Luis used 7 cubes. Is 7 more than 6? How do you know?

Kellen used 8 cubes. Is 8 more than 6? How do you know?

This understanding has been developing as students explore counting, the recognition that each number in the counting sequence is 1 more than the number previously said.

Have students take apart their "more" tower so they just have the original tower with 6 cubes.

Look at your tower. How many cubes are in your tower again?

Can you make another tower, this time with less, or fewer, cubes?

Observe them as they build.

How do you know this tower has fewer cubes than your other tower?

Ask students to count the number of cubes they have in their new tower. Ask them to share how many they used. Record the numbers on the board. Talk about each as you record it.

Bailey used 4 cubes. Is 4 less than 6? How do you know?

Colin used 2 cubes. Is 2 less than 6? How do you know?

Have students take apart all of their towers.

Now, I'd like you to build me a tower with 5 cubes, a tower with less than 5 cubes, and a tower with more than 5 cubes. Be ready to show me your towers and tell me about more or less.

Have students share their towers and discuss how they knew the towers were more or less.

Matching Bears SHOW IT!

Students explore the strategy of lining up and matching materials to compare them. Colored bear counters are used in this activity, but any colored manipulative will work.

Make a pile of red bears and a pile of blue bears (no more than 10 in a pile). Do not line them up; just put them in a pile. It should be hard to determine the quantity simply from looking at the piles.

Which pile has more bears?

Talk to your partner. What could you do to make it easier to figure out which pile has more bears?

Have students share their ideas. Some may suggest counting them (a good suggestion). Others may suggest lining them up (another good suggestion).

Take the suggestion of lining them up and line them up without starting at a common starting point or with uneven intervals.

Does this help? Why or why not?

What could we do so we don't get confused?

Listen to students' suggestions. If no one suggests lining them up so they are next to each other (matching), suggest it through a think-aloud.

Maybe if I started here and lined them up so they are right next to each other, then I could see which line has more bears.

Which has more now? How can you tell?

Show with your fingers how you matched the first red and blue bears, then the second red and blue bears, then the third red and blue bears.

Now it is easy to see which has more.

Which has less? How can you tell?

Common Error

Students often line up counters with different starting points or gaps between objects so the comparison is not accurate. Specifically addressing ways to line up items is essential for many students.

Give students zipper bags with two colors of bears and have them line them up, using matching to figure out which color has more or fewer bears (see photo).

This student lines up the bears to see which quantity is greater. Although she wrote the numbers, she was actually using the line of bears rather than the numbers to determine which was more. The next step will be determining which is greater by looking only at the numbers.

Just Enough Carrots

Students listen to *Just Enough Carrots* by Stuart J. Murphy and use the story context to explore more, less, and same.

Before Reading:

Ask students to turn to partners and discuss:

If I gave you 2 yummy pieces of candy, would you want to have more or fewer? Why?

If I gave you 3 smelly skunks, would you want to have more or fewer? Why?

What would you like to have more of?

What would you like to have fewer of?

Tell students you are going to read a story about more and fewer of things.

During Reading:

Read *Just Enough Carrots* to students.

While they are listening, students should be thinking whether they would like to have more or fewer of the items in the story.

After Reading:

Talk about same, fewer, and more.

What does it mean to have more of something?
What does it mean to have fewer of something?
What does it mean to have the same amount of something?

Give students fewer/same/more mats.

Give students counters to place on their mat showing same, fewer, and more.

We have 5 carrots. Show me 5 in the same box.

Have students count to be sure they have 5 in the same box.

Show me fewer than 5 in the fewer box.

Have students share how many they have in the fewer box, checking to be sure that it is less than 5.

Show me more than 5 in the more box.

Have students tell how many they put in the more box.

How do you know it is more?

Do more examples with counters, discussing students' responses to each.

Then, have students draw 6 of something (anything they'd like) in the Same section of their mat. Ask them to draw fewer than and more than 6 of that item in the other sections.

Review their drawings to assess their understanding of fewer and more.

If students have drawn too many or too few items, ask them to count the items to check their work.

Using a Balance to Compare

Students use a balance to compare numbers using greater than, less than, or equal to.

Students in kindergarten learn to use the phrases greater than, less than, and equal to but do not use the symbols >, <, or = to compare numbers.

Each group of students will need a classroom balance and two sets of 10 connecting cubes in different colors for a total of 20 cubes.

Today we are going to use a balance to compare numbers.
Put 6 red cubes in one side and 3 blue cubes in the other side.

Watch as students place cubes.

What do you notice about the balance?
Why do you think the side with 6 cubes goes down?
Why do you think the side with 3 cubes goes up?
Look at the side with 6 red cubes. It is down because there are more cubes in that side.
When there are more cubes, we say the number 6 is greater than the number 3.

If students are not familiar with a balance, allow them time to play with it and the cubes so they can explore what happens when cubes are placed in each side.

Add *greater than* to the word wall or Math Talk chart.

It is important when comparing that students accurately count the cubes.

Connect to measurement, Module 10.

How can you use greater than *to tell about the numbers 6 and 3?*

There are more cubes on the side with 6 so 6 is greater than 3.

Look at the side with 3 blue cubes.

It is up because there are fewer cubes in that side.

When there are fewer cubes, we say the number 3 is less than the number 6.

Add *less than* to the word wall or Math Talk chart.

How can you use less than *to tell about the numbers 3 and 6?*

There are fewer cubes on the side with 3 so 3 is less than 6.

Have students remove the cubes from the balance.

Let's compare two other numbers.

Put 4 red cubes in one side and 9 blue cubes in the other side.

What happened to the balance?

Why did the side with 9 cubes go down?

Point to the words *greater than* on the word wall.

Use greater than *to tell your partner about the numbers 9 and 4.*

Listen as students talk with their partner.

How did you use greater than *to compare 9 and 4?*

There are more blue cubes so 9 is greater than 4.

Point to *less than* on the word wall.

Use less than *to tell your partner about the numbers 4 and 9.*

Listen as students talk with their partner.

How did you use less than *to compare 4 and 9?*

There are fewer red cubes so 4 is less than 9.

Have students remove the cubes from the balance.

Let's compare two other numbers.

Put 7 red cubes in one side and 7 blue cubes in the other side.

What happened to the balance?

Why didn't the balance go up or down?

Add *equal to* on the word wall or Math Talk chart.

How can you use equal to *to tell about the numbers 7 and 7?*

Yes, 7 is equal to 7.

When there are the same number of cubes, we say the number 7 is equal to the number 7.

Students continue to show other numbers in the balance and use greater than, less than, or equal to as they compare the numbers.

Students will need ongoing practice to use greater than, less than, and equal to compare numbers.

Thinking Through a Lesson

Building Numbers on a Number Strip

Students build towers of cubes for each number on a 0–10 number strip to visualize that each counting number is 1 more than the number before it. They connect this idea to the concept of *more* as they transition to comparing quantities using numbers only.

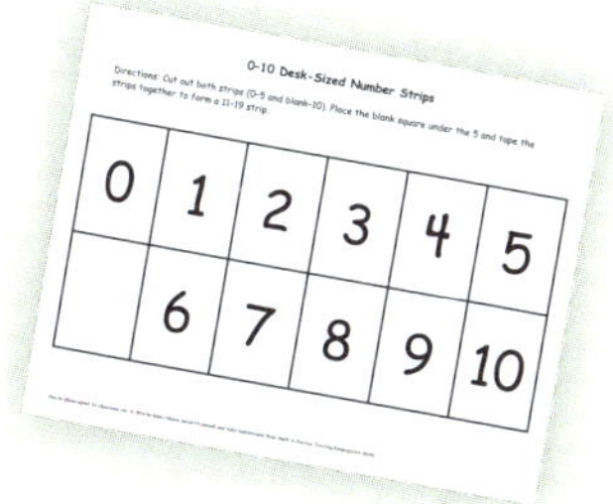

Give pairs a large number strip (0–10) and connecting cubes.

SMP4

We are going to make towers to show each of the numbers on your number strip.

SMP2

How many cubes should we put on the 0? Why? (None because 0 means nothing.)

How many cubes should we put on the 1? Why? (1 cube to show 1.)

How many cubes should we put on the 2? Why? (2 means 2 cubes.)

Make towers to show all of the numbers on your number strip.

These initial questions allow us to see whether students understand the directions. After a couple examples done together, observe as students then work on representing the remaining numbers with towers.

When they have finished, ask them to talk with their partners about what they notice about the number strip towers.

Then, have them share with the group.

SMP8

What do you notice? (The towers get bigger from 0–10.)

How much bigger? (They get 1 cube bigger for every number.)

Does that make sense? Why or why not? (Yes, because you have more every time you count another number.)

Listen for indications that students understand that each number is 1 more than the prior number. If they are simply saying they get bigger, ask for more precision with questions like, "What do you notice about the way they are getting bigger? What can you say that tells me exactly how much bigger the next number is?"

Pose some comparison problems and have them look at the towers to solve them.

Which is more, 6 or 3? Explain. (6 is bigger.)

Which is more, 2 or 5? Explain. (5 is bigger.)

Which is more, 1 or 4? Explain. (4 is bigger.)

Focus on the term more. Ask students to tell you how they know which is more. They may use words ("It's taller") or refer to the towers placed side by side as they explain.

What do you notice about the numbers that are more than the others? (They are farther down the number strip.)

Do you need the towers to figure out which has more? (No, if it is after a number when you count then it is more.)

Focus students on the order of the numbers on the number strip. If needed, go back to the previous examples that students compared using their towers and look at the number strip for each. Have students place a transparent counter on each number on the strip and discuss the location of each number on the strip (Which is farther from 0, 6 or 3?)

Remove the towers, but leave the number strip.

Which is more, 7 or 3? How do you know? (7 because it is farther down the strip.)

We say that 7 is greater than 3.

For students who need additional support, allow them to rebuild towers to visualize each number pair, then locate the numbers on the number strip. Have them place counters on the numbers on the strip and talk about which is more (or greater than) as they look at the placement of their counters.

This lesson bridges comparing sets of objects to comparing with numbers only. Students who are still having difficulty comparing sets of objects may need the support of counters or cubes as they move to the following lessons. Continue to connect the quantities in each set to a number line, or a hundred chart, to help students gain insight into how to compare numbers using their understanding of the counting sequence rather than counting the objects.

In previous activities, we compared how many cubes or bears, but now we are beginning to compare numbers. We say we have more cubes when we have 6 rather than 3, but when we talk about numbers, we say that 6 is greater than 3. We are comparing the values of the numbers. Use both words frequently as you compare: "There are more cubes in our tower of 6 than in our tower of 3. 6 is greater than 3."

SMP6

Record *greater than* on your word wall or Math Talk chart next to the word *more*.

Which is more, 3 or 5? How do you know? (5 because it is farther down the number strip.)

We say 5 is greater than 3.

Which is more, 1 or 6? How do you know? (6 because it is farther from 1 when we count.)

We say 6 is greater than 1.

The symbols for greater than and less than are not introduced until first grade. Students in kindergarten are getting comfortable with the language of more and greater than, fewer and less than.

In closing, have students turn and talk as you listen to their comments. Then, have some students share with the group.

How does using a number strip help you compare numbers? (You know that every number you say is 1 more, so if you look at the strip you can see which are more.)

How does knowing how to count help you compare numbers? (Because every time you say another number, it is 1 more.)

Variation: This lesson can be repeated with an emphasis on identifying towers with *fewer* cubes or the number that is *less than* the other.

These questions help determine which students have begun to see that the counting sequence helps them compare numbers. Many students require additional exposure to gain this insight.

FORMATIVE ASSESSMENT

Option 1

Show students a group of counters in random configurations, such as a group of 7 and a group of 5 counters. Ask:

> *Which group has more?*

They may touch and count each or line up the counters to compare.

Observe as the student works to count or arrange the counters.

Record observational notes of the student's work and responses.

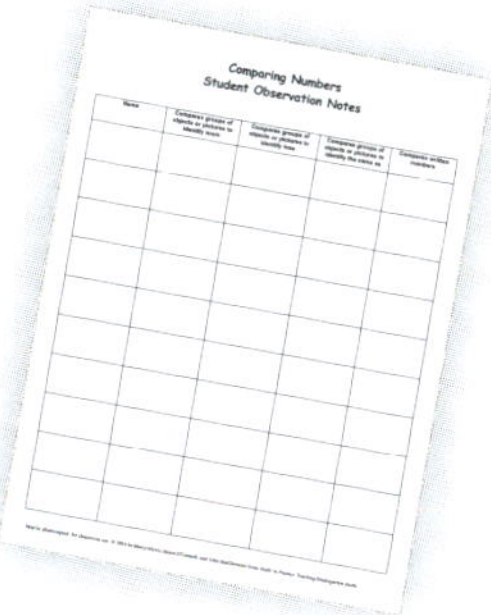

When the student responds that 7 is more, ask:

> *How do you know?* (Students might explain that 7 is more because when the cubes are lined up, 7 is longer to show more. Students might also have developed the insight that 7 is more because it is farther along in the counting sequence.)

Then ask:

> *Which group has less and why?* (Students might explain that 5 is less because when the cubes are lined up, 5 is shorter to show less. Students might also have developed the insight that 5 is less because it is earlier in the counting sequence.)

If students are not able to identify which group is more or less and they do not line up the cubes to compare, suggest lining them up.

Option 2

Provide students with pairs of numbers and the following task:

> Which number is greater?
>
> Draw a picture to show each number. Circle the number that is greater.

This student correctly drew pictures to show the numbers but circled all of the numbers on the page rather than circling the number that is greater in each set.

It's not clear if the student misunderstood the directions or could not identify the number that is greater. To reassess the student, have him count aloud the pictures he drew for each set of numbers. Tell the student to use a crayon to circle the number for the set that is greater. If needed, rephrase the question asking him to circle the group that has more items in it. If the student is not able to identify the set that is greater using the pictures, have the student use cubes to count and compare numbers.

Additional Ideas for Support and Practice

The following ideas extend students' understanding of comparing sets of objects and provide meaningful practice.

Cups of Counters

Students practice comparing quantities with counters in cups.

Put a different number of counters in each of two cups (e.g., 3 in one cup and 5 in the other cup). Have students dump the cups and decide which cup has more. Have them explain how they know.

Did anyone count to help them decide? How did counting help?
Did anyone match counters to help them decide? How did matching help?

Have students draw the counters in each cup and circle the drawing that shows more.

Repeat with two different cups.

Students work to compare numbers by lining up the counters. She has 5 counters and he has 2 counters. They will be able to see that 5 is greater than 2.

Scoop and Compare

Students practice comparing with an interactive task of scooping pom-pom balls.

Prepare bowls (or bags) of large pom-pom balls with a scoop or ladle for each pair of students. Students can take a handful if scoops are not available.

Player A scoops out pom-poms and counts them. Player A will say, "I have ___ pom-poms."

Player B scoops out pom-poms and counts them. Player B will say, "I have ___ pom-poms."

Players talk and decide who has more or less pom-poms, then spin the more/less spinner to see if the player with more or less wins!

Students place those pom-poms aside and scoop again. Continue play until all of the pom-poms are used.

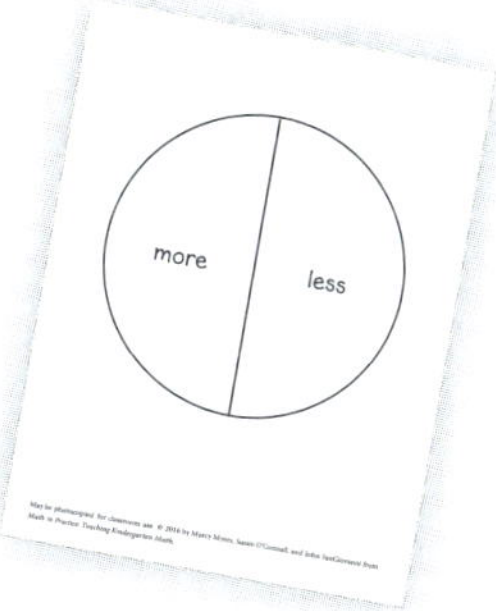

Talk About It/Write About It

Jenny lined up counters to see which group had more. Was this a good way to line them up? Explain.

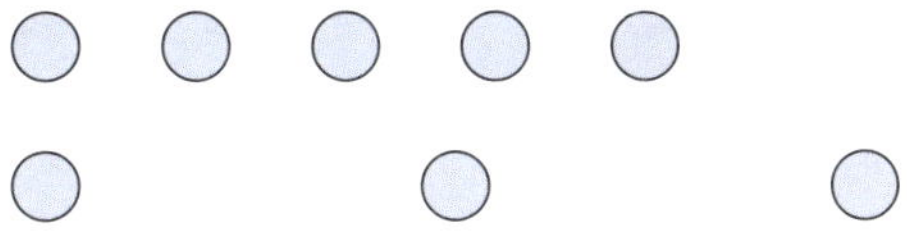

Ali had 4 crayons. Jamie had more than that. Draw a picture to show how many Jamie had. Label your picture with the number she had. Be ready to share. Is there more than one answer?

Draw a picture to show 8 is greater than 5.

Draw a picture to show 3 is less than 4.

Draw 6 balloons. Draw fewer than 6 balloons. Label the number of balloons you drew.

Draw 7 beach balls. Draw more than 7 beach balls. Label the number of beach balls you drew.

Count and Compare Center

This center provides counting practice and practice writing numbers, as well as practice comparing numbers.

Place counters, tiles, bears, cubes, and so on of two different colors into bags.

Students dump out the materials in each bag, compare them by matching and counting. Students then record the numbers using that color crayon (red for red counters or blue for blue counters), and circle the number that is more.

Students then replace the materials in the bags and pick another pair of bags.

Having the students record their work in colors allows you to assess the accuracy of their counting because you know there were 5 green bears and 4 red bears or 6 purple cubes and 3 red cubes.

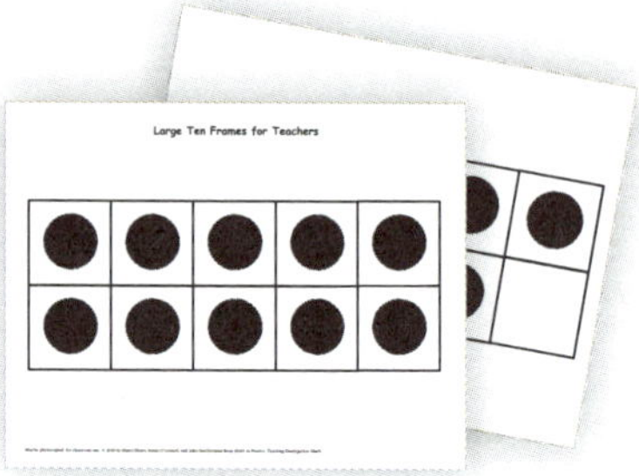

One More Fast Flash

Show students dots on a ten frame, and have them say the number that is 1 more.

When students have had ample experience with Fast Flash (see page 46), display a ten frame with 8 dots.

What number is 1 more than the number shown? (9.)

Add 1 more dot to the ten frame using another color to show 9.

Talk about how 9 is 1 more than 8, helping students understand that the next number in the counting sequence is 1 more.

Continue with other numbers showing 1 more.

Eventually, students should be able to say the number that is 1 more when the ten frame is shown.

In their initial experiences, students compared sets of objects. Then they compared objects with number cards visible to show the quantity in each set. Numbers were recorded, but the objects were still present. Finally, the objects are removed and students compared quantities with numbers.

COMPARING WRITTEN NUMERALS 1–10

Students have investigated comparing quantities with objects and drawings and are now transitioning to comparing numbers looking only at the numerals.

Number Line Circles SHOW IT!

Students circle numbers on a number line to see which is closer to or farther from 0.

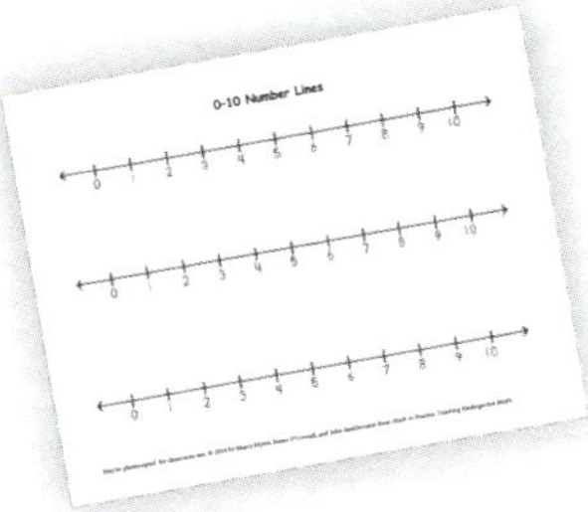

Give students two bags of counters with 1–10 counters in each bag.

Have them count the number of counters in each bag and circle the numbers on a number line.

Which bag has more counters? How do you know?

Were you able to tell which had more by looking at the number line? How?

How do you know that the number of counters that is farther to the right is more?

Add some ideas using a think-aloud method if no one suggests it.

I was thinking the numbers get smaller the closer I get to 0.

I was thinking the numbers get larger the closer I get to 10.

I was thinking that the bigger number is farther from 0. What do you think about that? Talk with your partner about it.

To bring the lesson to a close, use the laminated number lines and dry erase markers.

Your first 2 numbers are 5 and 2.

Draw a circle around the number that is more and underline the number that is less. Explain.

Repeat with additional sets of numbers.

Use a laminated number line or place the number line in a transparent pocket so students can circle the numbers, make their decisions, and then wipe off the circles to start again. Record some of their ideas on the board.

Differentiation

Some students may need to compare the actual objects longer than others. Allow them to use materials, but also have them select number cards or write numbers to show each comparison.

Target Number

Students use a target number and their understanding of more and less to compare quantities. This activity works well with a whole class.

Shuffle a set of large 0–10 cards (or use number cards displayed on an interactive whiteboard).

We have compared numbers using pictures and objects. Today we are going to compare numbers using number cards.

What words or phrases do we use when we compare numbers?

We are going to use greater than and less than to compare numbers to a Target Number.

Write *Target Number* on the board. To the left, write *Less Than*. To the right, write *Greater Than*.

Draw a card from the deck.

What number do I have? (6.)

6 is our Target Number.

Place 6 on the board under *Target Number*.

Have a student come up and select the next card.

Bobby picked the number 9. Our Target Number is 6. Is 9 greater than or less than 6? Where does 9 go on the board? Turn and talk to your partner and be ready to tell me where to place it and why.

Listen as students share with partners, and then have a student tell where it should go and why.

Do you agree? Does anyone have a different idea?

Allow students to question and share their ideas about whether 9 is greater than or less than 6.

Refer to the class number line or hundred chart to help students see that 9 comes after 6 when counting so it is greater than 6.

Let's pick another number card.

Have a student come up and select the next card.

Sarah picked the number 3. Is 3 greater than or less than our target number 6? Where does 3 go on the board? Talk to your partner and be ready to tell me where to place it and why.

Allow students to question and share their ideas.

Why is 3 less than 6? How do you know?

Continue having students draw cards from the deck and sorting as greater than or less than the Target Number.

This activity can continue over several days with different Target Numbers.

When 10 is the Target Number, ask students why all numbers in the deck are less than 10. What numbers could you write that are greater than 10?

This activity can become a center using smaller number cards and the Target Number recording sheet.

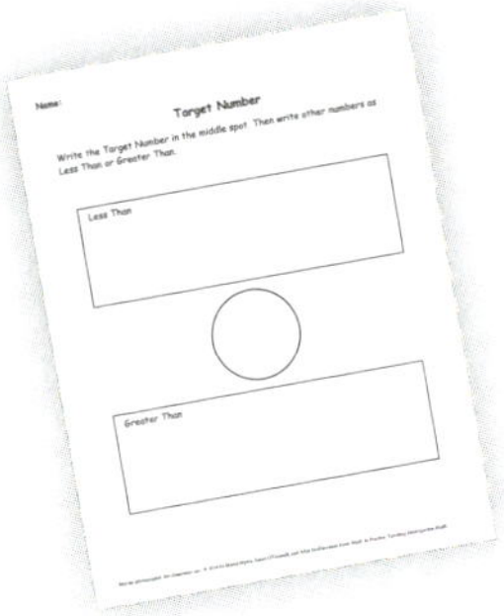

Students draw a card and record the number in the Target Number Spot. As they pick numbers, they write the numbers in the Less Than box or Greater Than box.

> **FORMATIVE ASSESSMENT**
>
> Students are given pairs of numbers cards.
>
> They hold the greater number up in the air and the number that is less is held lower. (For example, give students cards with 8 and 3 on them. Students will hold up the number 8 and the number 3 will be lower.)
>
> Record student responses on the Observational Notes page.
>
> If students are not able to compare the quantities by looking at the numbers, put out counters to show the numbers 8 and 3.
>
> Then ask which number is greater and which number is less.
>
> Have students tell you why 8 is greater and 3 is less to assess whether they understand what *greater* and *less* mean.

Additional Ideas for Support and Practice

The following ideas extend students' understanding of comparing numbers to 10 and provide meaningful practice.

Talk About It/Write About It

Use the following as discussion starters:

Jay has 6 apples. Does he have more than 5? How do you know?
Maddie says 9 stickers is more than 7. Is she right? How do you know?

Students can draw or write to show that one number is more than or less than another.

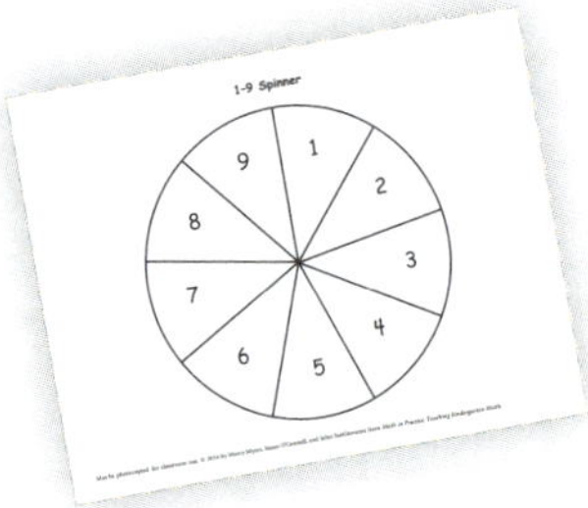

Spin and Show 1 More

Partners practice showing 1 more.

Materials: 1–9 spinner for the pair; a ten frame and two colors of counters for each student

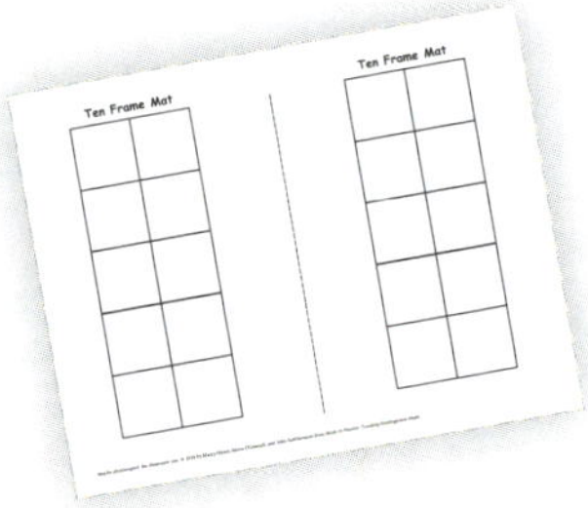

Player A spins the spinner and uses one color of counters to show the number he spun on the ten frame. He says, "I have 6; 1 more is 7," and places 1 counter of a different color on his ten frame.

Player B takes a turn following the same process.

After both players have shown their number, they compare frames. The player with the higher number wins that round.

Continue play.

Compare the Cards

This small-group activity provides practice in comparing numbers.

Have a stack of 1–10 cards.

Pull two cards.

Which is more/greater? How do you know?

Which is less? How do you know?

Allow students to use a number line, if needed.

Ask students to draw a model of each number to justify their answer.

Roll and Compare

This center or partner practice activity involves rolling dice and comparing numbers.

Students work with partners and take turns rolling a 1–6 number cube or a 0–9 ten-sided die (dependent on students' level of experience).

Partners record the numbers and a model of each die or cube in their math journals or on paper, then circle the number that is less (or more, whatever you designate).

Then, players roll and start again.

Which Number Is Greater?

This two-player interactive game can be used in centers or for partner practice. It is similar to the traditional card game War.

Materials: two sets of number cards 0–10 (22 cards in the deck)

Shuffle the deck and place it facedown between the students.

Player A draws a card and turns it over.

Player B draws a card and turns it over.

The player with the greater number says, "_____ is greater than ____," and takes the two cards.

If the cards turned over are equal, players can turn another set of numbers to determine who gets the cards.

Play continues until all cards are drawn from the deck.

The player with the most cards is the winner. Then cards are shuffled and the game starts again.

Vocabulary

balance	*greater than*
compare	*less than*
count	*match*
equal to	*more than*
fewer than	*same as*

ONLINE RESOURCES

General Resources

Additional Problems

Comparing Numbers Student Observation Notes

Large Ten Frames for Teachers

Number Cards 0–10

Number Lines 0–10

Number Strip

Ten Frame Mats

Resources for Specific Activities

1–9 Spinner

Fewer/Same/More Mat Template

More/Less Spinner

Target Number Recording Sheet

Which Number Is Greater? Sheet

These resources are available at http://hein.pub/MathinPractice, keycode MIPGK.

MODULE 6

Decomposing Numbers

About the Math K.OA.1; K.OA.3; K.OA.4

One of the big ideas about our number system is that numbers are flexible. They can be broken apart (decomposed) and put together (composed): 6 can be decomposed into 5 and 1, 4 and 2, and 3 and 3. Although kindergarten students explore this idea at a simple level, decomposing numbers continues to resurface as students move from grade level to grade level. In first grade, they decompose 27 into 2 tens and 7 ones or 20 and 7. They also realize that 27 can be decomposed to 1 ten and 17 ones, just another of the many ways to break apart the quantity of 27. Second graders decompose 348 into hundreds, tens, and ones. Third graders decompose to add and subtract three-digit numbers by place value. Fourth graders decompose fractions and decimals. The concept of numbers being composed and decomposed is central to understanding the way numbers work.

The key ideas focused on in this module include:

- **decomposing numbers to 10 using manipulatives, pictures, diagrams (e.g., part-part-whole diagrams), and equations**
- **exploring decomposing 10 in greater depth, including when given one number, finding the other number that makes 10**
- **using decomposing numbers to solve math problems.**

Explorations with decompositions in kindergarten are active and engaging.

- Students put 8 two-color counters in a cup and then shake and spill the counters to find different combinations of red and yellow counters (see Figure 6.1).

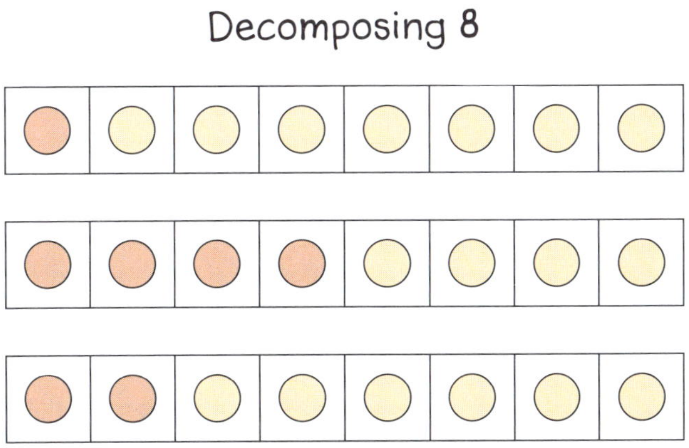

Figure 6.1 Students spill 8 two-color counters and use frames to visualize the decompositions.

- They use bead counters with 8 beads to slide some left and find out how many are to the right.
- They use chains of 8 connecting cubes, breaking the chain into 2 parts to find the decompositions.
- They draw pictures of 8 balloons and color them to show different quantities of red and blue balloons.
- They represent their models with equations (8 = 6 + 2).

Exploring the decompositions for numbers 10 or less, recording the decompositions with drawings and numbers, and discussing their insights build our students' understanding of numbers.

In addition to gaining insight about how numbers work, composing and decomposing numbers forms the foundation for addition (joining two quantities) and subtraction (separating a quantity). As students push two sets of counters together to form a new group or connect two chains of cubes to build a new chain, they are seeing the action of joining that they are later introduced to as addition. As they break chains into two pieces or split beads into two parts, students see the action of separating that will later be introduced as subtraction. Showing decompositions as addition equations helps make this connection for our students (see Figure 6.2).

Figure 6.2 Using equations to express the decompositions helps our students connect this action to addition.

Number bonds (see Figure 6.3) provide students with a way to show the two parts and the whole (total). The number bonds can be drawn in different arrangements to

Learning Goals

I know that numbers can be broken into parts.

I can break numbers apart using objects, drawing pictures, or showing part-part-whole.

I can write equations to show how I break numbers apart.

I can find two numbers that equal 10.

help the students think flexibly about representing numbers. Number bonds provide a visual foundation for writing equations.

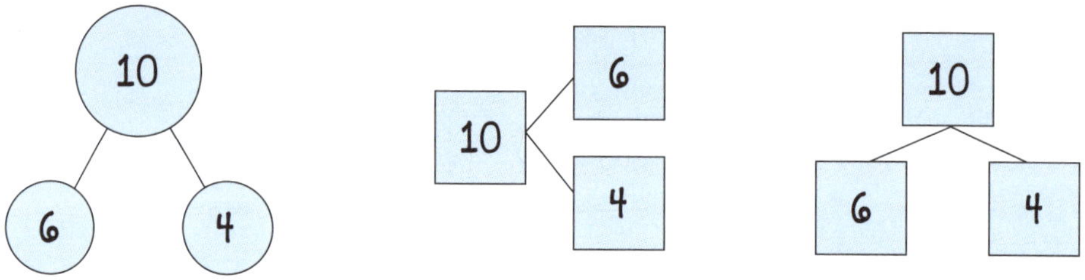

Figure 6.3 The number bonds may be positioned in different ways, but both parts are always connected to the whole. Knowing which number represents the whole and which numbers represent the parts is an important understanding.

Exploring the addends that form 10 is particularly important. The understanding, and memory, of addends that combine to make 10 allows students to do mental computations more easily. At the kindergarten level, the focus is on finding the other addend that makes 10 when posed a number 1–9 (see Figure 6.4), not on knowing these from memory. First graders will then work on committing these addend pairs to memory.

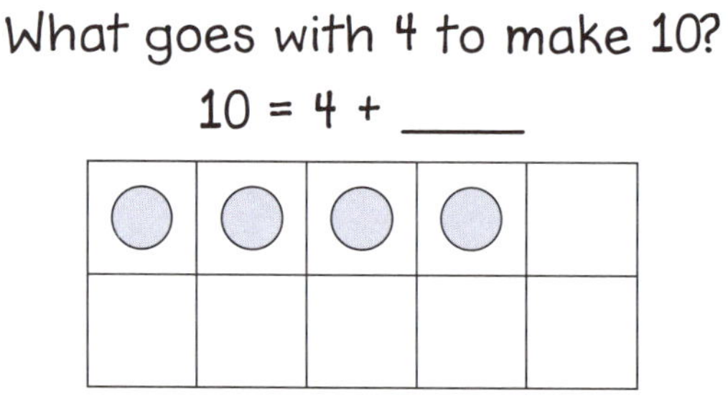

Figure 6.4 Students find the other addend that makes 10 through explorations with objects and drawings.

See connections to subtraction and problem solving in Module 8.

The ability to decompose numbers is revisited and extended as students explore math problems that require them to separate numbers into sets. Connecting those problem-solving experiences to these initial experiences in which students decompose sets of counters or cubes provides them with a foundation for problem-solving tasks. Consider the problems:

- There are 5 cats. Some are black and some are white. How many of each?
- There are 6 lollipops. Some are lemon and some are cherry. How many of each?

In each case, students explore a context in which they must find possible ways to separate (break apart) the total number of cats or lollipops into parts. They are utilizing the decomposing skills they explored with objects, drawings, part-part-whole charts, and equations.

Exploring the Progression

PREVIOUS	NOW	NEXT
Preschool	**Grade K**	**Grade 1**
Probably no experience with composing or decomposing numbers, only counting total objects	Decomposing numbers to 10; using decomposition to solve math problems; writing equations to show decomposition	Decomposing and composing numbers to separate tens and ones in varied ways to add or subtract (12 + 7 = 10 + 2 + 7); memorizing addend combinations of 10

Ideas for Instruction and Assessment

USING OBJECTS AND DIAGRAMS TO DECOMPOSE NUMBERS

Hands-on tasks help students visualize the decomposition of numbers. Experiences with connecting cubes, linking chains, two-sided counters, and other manipulatives provide students with opportunities to see how numbers can be decomposed in different ways, and diagrams and charts help them visualize the part-part-whole relationship. As their understanding increases, students explore representing their decompositions with equations.

Breaking the Chain SHOW IT!

Students build chains of connecting cubes and then break the chains into two parts to explore decomposing numbers. The terms *part* and *whole* are used in context throughout the lesson as students explore and draw the whole chain, as well as the two parts once the chain is broken.

Give each student 10 connecting cubes and a small whiteboard (or paper) for recording.

Build a chain of 5 connecting cubes.

Compare your chain to your partner's chain. Are they the same size? Do they both have 5 cubes?

Break your chain into two parts.

How many cubes are in one part? How many cubes are in the other part?

Draw your new chains on your whiteboard.

Write the number of cubes by each part.

Have students show their drawings and share how many cubes are in each part of their chain.

Record some of the drawings, labeled with numbers, on the board.

Did everyone's parts look the same? Tell me what you notice.

Lessons in This Module

Make sure that students don't confuse whole with hole.

The words part and whole are used in context throughout the lesson.

Students' skills at counting and one-to-one correspondence as well as their ability to write numbers all play a part in their ability to perform this math task.

Have students put their two parts back together to make the original chain.

What do you notice? How many cubes are in your whole chain?

Does that make sense? Why?

But Colin put together 3 and 2, and Liam put together 4 and 1.

> Rather than telling students, we are asking questions to get students to notice that there are multiple ways to break numbers apart.

Have students erase the whiteboards and break apart their chain into single cubes.

Now, build a chain with 7 cubes. The whole chain has 7 cubes.

Break your chain into two parts.

Draw your two parts on your whiteboard and write the number of cubes by each part.

Turn and tell your partner about your whole chain and the two parts you made. Be sure to use numbers when you talk about your chain and the parts.

After students have shared with partners, ask some students (with different decompositions) to share their parts as you record them on the board with drawings labeled with numbers.

What do you notice about the parts when we broke chains of 7?

Have students put their two parts back together to make the original chain.

What do you notice? How many cubes are in your whole chain?

Does that make sense? Why?

For practice, have students build and draw a whole chain of 8 cubes and label it.

Under that drawing, have them draw at least two ways the chain of 8 cubes can be broken into two parts and label the parts.

> As students work, look to make sure they are working with 8 cubes and that 8 has been decomposed and labeled accurately.

FORMATIVE ASSESSMENT

Give students 5 connecting cubes and ask them to make a chain.

Tell them to break the chain to show 5 in two different ways.

Have them record each way by drawing the two parts and recording the number of cubes in each part.

Chips, Chips, What Color Chips? SHOW IT!

Students use a specified quantity of 2-sided counters (the whole) and spill them from a cup, noticing the number of counters that are yellow or red (the parts). Students begin by labeling the two parts with numbers, but progress to recording the parts on part-part-whole mats and then to equations. Doing this lesson in a small group will allow all students to see the counters.

> In some 2-color counter sets, the counters are red and white rather than red and yellow.
>
> Some teachers make handmade counters by using flat, dry beans (like kidney beans), and spray-painting one side of the beans.

> The speed at which you move through these representations of the parts and whole depends on your students. Spend more time at any point if students need additional time to discuss and practice.

Show students some 2-color counters.

What do you notice about my counters? (One side is red and the other is yellow.)

What will happen if I spill them from a cup? What will you see?

Have students count 6 counters with you. Put the counters in a cup and spill them onto the table.

What do you see? Tell me using numbers.

Record the number of yellow and red counters on the board.

Do we still have 6 counters if some are red and some are yellow?

How can you prove there are still 6 counters?

> We are asking students to be precise in their explanations by prompting them to use numbers as they talk about what they see.

Write the addition equation on the board (e.g., $6 = 4 + 2$) saying,

6 is the same as 4 + 2.

Do you think we will always get 4 red and 2 yellow when we spill our 6 counters?

Put 6 counters in your cup and try it.

Count how many red and yellow and be ready to share.

Allow students to share their ideas as you record on the board.

What happened? Was it 4 red and 2 yellow again?

They should be excited to tell you what they saw when they spilled their counters; for some, it may have been 4 and 2, but for others it was not.

> Although we are recording the parts and whole as an equation, we are not placing an emphasis on the equation at this time. Students are simply being exposed to this as a way of recording the parts and whole.

As you record equations, restate them using the context of the counters. As you record $6 = 3 + 3$, say,

So, 6 counters is the same as 3 red counters and 3 yellow counters.

As you record $6 = 1 + 5$, say,

So, 6 counters is the same as 1 red counter and 5 yellow counters.

> Students may notice that 1 + 5 and 5 + 1 are both decompositions. In this context they are not the same because 1 red and 5 yellow was a different spill than 5 red and 1 yellow. Although we don't really stress the commutative property at this level, an important insight for our students is that the whole is the same (6) when the parts are 1 and 5.

As you record their data, ask questions like these:

Are you sure you still have 6 counters altogether?

How can you prove there are still 6 counters?

Ask if any students had all red counters (or all yellow counters). If a student had all red counters, you might ask,

So, how can I show how many yellow counters? How many were there?

How do I show none?

Record $6 = 6 + 0$, saying,

So, 6 counters is the same as 6 red counters and 0 yellow counters.

> Addressing 0 in equations should happen naturally as it occurs in problems and investigations. Having 0 yellow counters is a great opportunity to review the use of 0 to stand for none.

Ask students to turn and talk to a partner:

Look at what I wrote on the board (all of the equations). *What do you notice?*

Why does each of these (equations) *begin with 6?*

Is there just 1 way to show 6 with these counters?

Does that make sense? Explain.

If time allows, give students 7 two-color counters and find different possibilities for breaking apart 7.

Showing Parts Through Pictures SHOW IT!

Students draw math pictures to show the way they break apart numbers. They discuss ways to show the parts through their drawings to build a foundation for part-part-whole diagrams.

Common Error

Students need to correctly represent the number to be decomposed or the "whole." If asked to represent the number 5 and they draw 6 circles by mistake, the parts of 5 will be incorrect.

Differentiation

Some may be able to make a connection to previous work with manipulatives and be able to draw 3 circles and then 2 circles to show 5. Other students might need to separate the number by drawing 1 circle on one side and then 1 circle on the other, alternating between the sides.

Select students who have shown different ways of drawing the parts or whose solutions are different.

Differentiation

Some students may benefit from the direct connection between manipulatives and drawing a picture. It could be helpful to have a train of 5 connecting cubes and ask the student to break apart the train. They can then draw a picture of the cubes to show decomposing 5.

For this small-group lesson, prepare individual dry erase boards and markers for each student to draw pictures.

You are getting very good at breaking apart numbers using manipulatives.

We have used counters, connecting cubes, and other manipulatives to show ways to decompose numbers.

What if you didn't have any counters or cubes? How could you show how to break numbers apart?

Responses will likely include that you could draw pictures.

Draw a picture to show me how you might break 5 into parts. Be sure I can see the two parts.

As students work, observe what they draw.

After students have worked, have them share and explain their drawing to a partner, telling their partner about the parts.

Select some students to share their drawing and their thinking.

I see that some of you drew real things, like balloons, and others drew circles or Xs. Is it okay just to draw circles or Xs?

How did you draw a picture to show the parts of 5?

Share some ways students organized their drawings as they decomposed numbers.

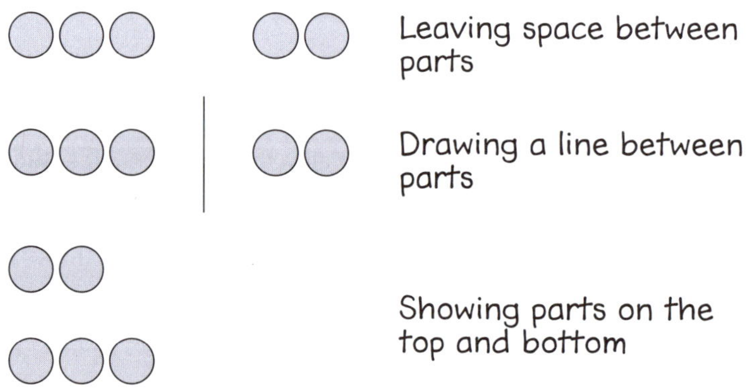

How do these drawings help you see the parts?

Ask students to turn and share a way they will break apart 6 and show the parts.

Have them do it and discuss their drawings.

FORMATIVE ASSESSMENT

Ask students to draw a picture to show different ways to decompose, or break apart, the number 4.

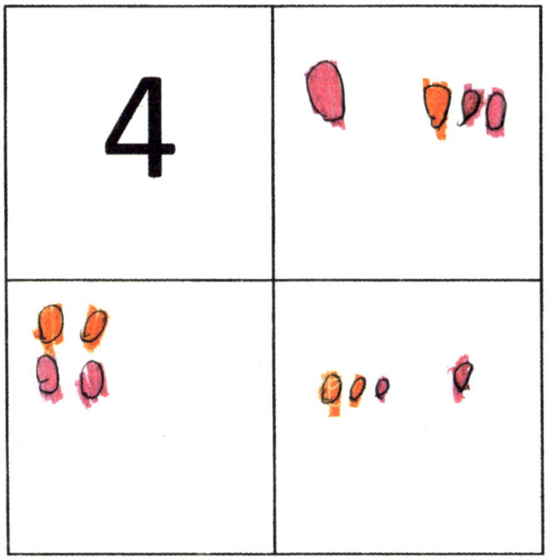

The student on the top decomposes the number 4 correctly using the placement of the circles to show 4 separated into 1 and 3, 2 and 2, and 3 and 1. The colors the student uses for the circles make her work a bit confusing, but when asked to explain it, she is able to tell about each decomposition. The number labels on the work on the bottom make that student's decompositions clear.

Modeling Using Part-Part-Whole Mats SHOW IT!

Placing manipulatives on part-part-whole mats helps students visualize decompositions. These mats then lead to using number bonds to record the part-part-whole relationship.

Divided paper plates with three sections also work well for part-part-whole activities.

For this small-group lesson, give each student a part-part-whole mat and a set of 10 manipulatives (e.g., counters, cubes).

Draw a larger part-part-whole mat on the board or on chart paper. Use disk magnets or small sticky notes to model student work as needed.

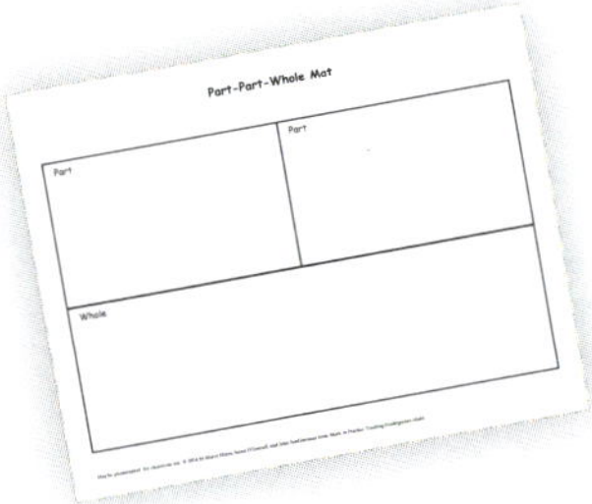

Give each student a part-part-whole mat.

> *What do you notice about the mat?* (It has two smaller sections and one big section. *Or,* There are three sections on the mat.)

Place 5 counters in the bottom section. As students watch, separate the 5 counters into 3 and 2 and move them to the smaller sections.

> *Talk with a partner: What did I just do?*
>
> *Why did I move the counters to the two smaller sections of the mat?*
>
> *How many counters are on the mat?*

Rather than telling students about the sections of the mat, we are asking them to look at the mat and think about those sections as we show some examples.

Have them watch again as you put 5 counters in the bottom section and then separate them into 1 and 4, putting 1 in the left section and 4 in the right section.

> *Talk with a partner: What did I just do?*
>
> *Why did I move the counters to the two smaller sections of the mat?*
>
> *What do you think the big section shows?*
>
> *What do the small sections show?*

Listen to see if students are connecting their previous work with parts and whole to explain how their mat is being used. Do they see the whole on the large section and the parts in the two smaller sections above? Do they see that the two parts equal the whole?

For this activity, we mention the whole section on the bottom and the parts on the top. The mat does not have to be designed in that way. The parts can be on the bottom and whole on the top, as long as students see the differences between the parts and whole on the mat they are using.

Have students show 6 on their mats.

> *Show me 6. Where would you place the counters on your mat to show the whole (6)?*

Watch as students model 6 on their mats.

> *Can you show me a way to break 6 apart by moving your counters?*
>
> *Be ready to explain what you did.*

Watch as they separate the 6 counters into two parts and show it on their mats. If students need help getting started, ask:

> *How could you move the counters to show the parts of 6?*

Have them explain what they did, including why they placed the parts in the two smaller sections of the mat.

Put your 6 counters back together. Where should we put them when we are looking at the whole 6? Why?

Break your 6 apart another way. Show me using your mat.

Have students share their parts and explain why they moved the parts to the smaller sections of the mat.

What can you see when we use this mat?

How could this help you?

What could we call this mat?

Let them be creative, but suggest *part-part-whole mat* if no one comes up with it, because the mat helps us see the parts and the whole.

Continue using part-part-whole mats over several days with a variety of manipulatives to practice decomposing different numbers.

Thinking Through a Lesson

Introducing Number Bonds

Students have decomposed numbers using manipulatives, drawing pictures, and using part-part-whole mats and now are introduced to number bonds to represent the part-part-whole relationship.

The lesson begins with a review of decomposing numbers using pictures.

Students use individual dry erase boards with markers to show their work.

Draw a picture on your board to show a way you could break apart, or decompose, 6. SMP4

Observe as students work.

Turn and tell your partner how you broke apart the number 6. SMP7

Have students share some possibilities with the class by holding up their boards.

Paige drew hearts on a part-part-whole mat to show that 6 can be broken into 4 and 2. SMP4

Some students may draw a part-part-whole mat and others may draw pictures of cubes, counters, or shapes.

Rather than taking the time for all students to redraw their picture on the class board, using individual dry erase boards allows them to share their work without taking up too much class time.

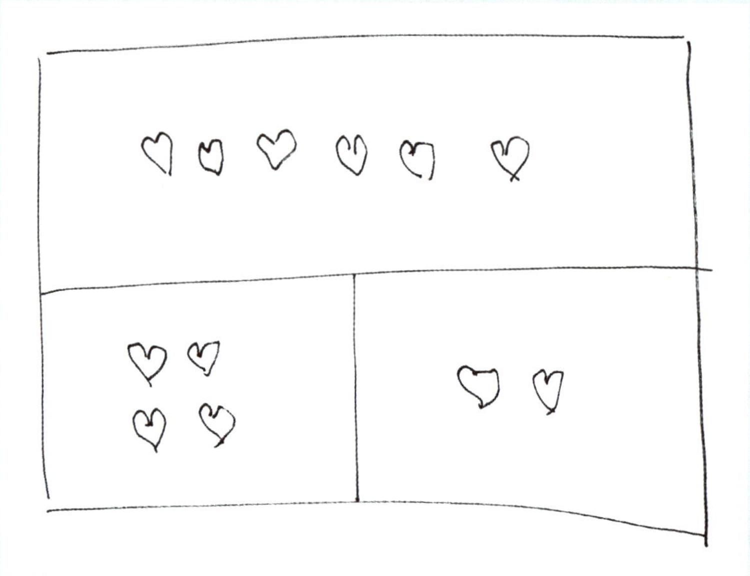

Record a number bond for 6, 4, 2 on the board.

Students have not seen a number bond before. They will be asked to figure out what it shows after seeing several examples. This gets them involved in the thinking and challenges them to make sense of the diagram.

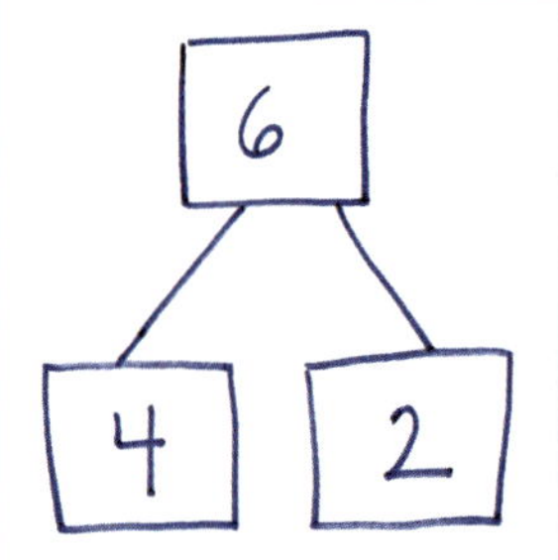

Kellen made circles on his board to show 6 as 5 and 1.

Record a number bond for 6, 5, 1 on the board.

Brendan drew a chain of cubes to show 6 as 3 and 3.

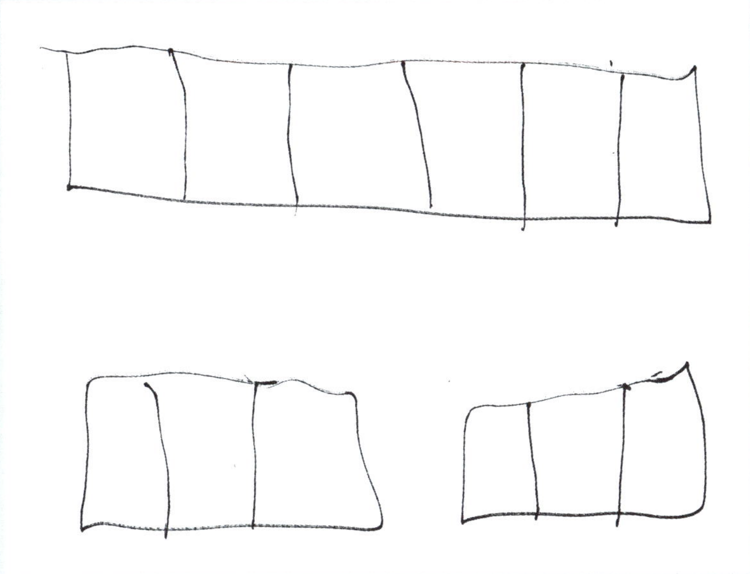

Record a number bond for 6, 3, 3 on the board.

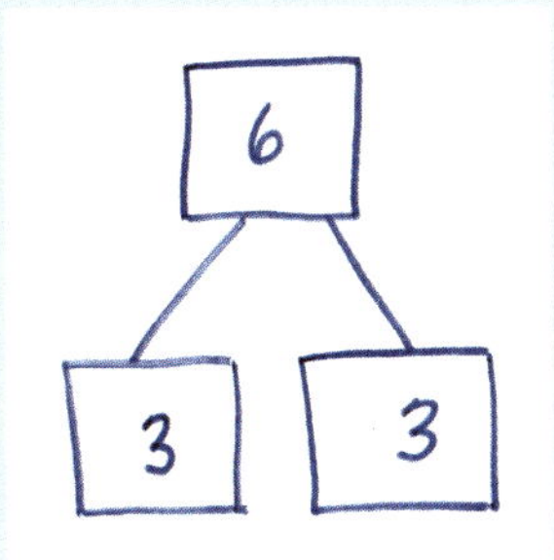

Students are asked to observe the new diagrams and try to figure out what they mean. Turning and talking to partners allows students to try out their ideas before they share with the class.

SMP8

Turn and talk to your partner: What did I just do? What do you think these numbers show?

Are students noticing that the parts and whole are on the number bonds? Are they noticing that the whole (total) is connected to the parts with lines?

Listen to the partner conversations.

What are the numbers? (Parts and the whole. *Or,* The numbers in our pictures.)

How is this like a part-part-whole mat? (It shows parts and the whole like on the mats.)

What are these lines? (They connect the parts to the whole.)

We just showed ways to break apart 6 without pictures of objects. We used a number bond.

Students figured out the diagram. We are naming it here. Creating an anchor chart or recording the term on a word wall or Math Talk chart helps remind students of this new term.

Ask the students to draw a number bond on their dry erase board to match their picture.

Have students share their number bonds with a partner, telling their partner what each number represents.

This allows you to see which students get the idea and which may still be confused.

Then, have a few students share with the class.

Ask students to decompose 7.

On one side of your dry erase board, draw a picture to show a way you could break apart 7.

SMP2

After you draw your picture, draw a number bond to show it with numbers.

Observe as students work.

How did you decompose 7? Show your partner your picture and your number bond.

Drawing the number bonds on the board as you talk helps support students who may need to see more examples.

Draw some of the number bonds on the board to show 7 is 1 and 6, 2 and 5, or 3 and 4.

Look at these number bonds. How are "7 is 4 and 3" and "7 is 3 and 4" alike? (They use the same numbers.)

How are they different? (The 4 and 3 are on different sides.)

SMP4

How could we show this with pictures? Or counters? (On a part-part-whole mat we could put 3 counters in one part and 4 counters in the other part or we could put the 4 and 3 on the other sides.)

Are students seeing that the 4 and 3 can be in a different order on the number bond? Can they connect this to what happens when they break apart 7 with concrete materials? They might have 3 and 4 or 4 and 3 but it still makes 7.

Have students practice creating number bonds to show the decomposition of several more numbers.

Turn and tell your partner what a number bond shows.

Listen for students to share the idea that number bonds show the parts and whole using numbers. They may say they are like part-part-whole mats, but with numbers instead of pictures or counters.

Differentiation

As students practice creating number bonds, invite the students who need additional support to join you for more practice. Begin with breaking apart the numbers with concrete materials, then revisit the creation of number bonds. Pose a number and, if needed, ask guiding questions like: "Where would you put the number that shows the whole (total)? Where would you put the numbers that show the parts?"

Quack and Count

In *Quack and Count* by Keith Baker, 7 ducks have fun playing peekaboo, chasing bees, and splashing and the illustrations and words provide a context for decomposing 7.

Before Reading:

Tell the students that you will be reading a book about 7 ducks.

Have them show a way to decompose 7 with their fingers.

Give each student a dry erase board and marker to record a number bond for each part of the story.

During Reading:

As the ducks are shown in arrangements that equal 7, students draw the number bond on their boards.

As students hold up their boards, record each number bond on the board.

After Reading:

Revisit the pictures and the number bond that represents each.

How could we write this as a number sentence?

If you are looking at the picture and number bond for 5 and 2, write $7 = 5 + 2$ on the board.

Ask students to find and record ways that 7 can be represented as number sentences (e.g., $7 = 7 + 0, 7 = 6 + 1, 7 = 5 + 2, 7 = 4 + 3, 7 = 3 + 4, 7 = 2 + 5, 7 = 1 + 6$ and, $7 = 0 + 7$).

After reviewing the number sentences, give the students 2-color counters.

What if there were 8 ducks?
Use your counters to find different ways to show the 8 ducks.
Write a number bond and a number sentence for each.

Additional Ideas for Support and Practice

The following ideas extend students' understanding of decomposing numbers and provide meaningful practice.

Finger Flash

While reading a counting book, students use their fingers to show decomposing numbers into parts.

Choose a favorite counting book to reread.

As you progress through the book, ask students to show and explain how they represent numbers on each page. For example:

I showed 7 using 4 fingers on one hand and 3 on the other. *Or*, I showed 7 using 5 fingers on one hand and 2 on the other.

Have students turn and talk to a partner to share their examples.

Some possible books:

- *Mouse Count* by Ellen Stoll Walsh
- *One Watermelon Seed* by Celia Lottridge
- *Dinner at the Panda Palace* by Stephanie Calmenson

Number Bracelets

Class sets of number bracelets can be made using pipe cleaners and beads. When a student puts a bracelet on her wrist, she can break apart the total number of beads showing some beads on the outside and some beads on the inside of the wrist (e.g., a student wearing the number 6 bracelet can show 6 as 1 on the inside and 5 on the outside or 2 on the inside and 4 on the outside).

To make a number bracelet: Place 2 beads on a pipe cleaner, then form a circle and twist the ends together.

Continue with numbers 3–10, using the same color for all beads on each bracelet.

It is helpful if the beads are a different color from the pipe cleaner.

To label the number of beads on each bracelet, create a tag by folding a piece of masking tape or adhesive label onto itself around a section of the pipe cleaner.

Label the amount on the tag with a permanent marker.

The label shows the total number of beads and students then separate the beads to show two parts. Students find multiple ways to decompose the number.

Capture and Count

SHOW IT!

Students use a cup and cubes to explore ways to decompose a number.

The teacher designates a number to be decomposed.

Have students place that number of cubes on the table.

Player A captures (covers) some of the cubes with a cup.

Then Player B counts the cubes captured under the cup and the cubes showing outside the cup.

Player A records the parts and whole with drawings and numbers. Students might record using a number bond or an equation to show the parts and whole.

Students begin again, capturing a different number of cubes and switching partner roles.

Shake a Bag

Students shake clear plastic bags to see how many counters are on each side of a line drawn down the center of the bag. Then, they record the parts using a number bond.

Prepare sandwich-size, resealable plastic bags by drawing a line down the center of the bag, from top to bottom, with a permanent marker, creating two sides of the bag.

Put a different quantity (2–10) of the same-color counters or beans in each bag.

Label the bag with the number of counters or beans.

Students shake the bags and then lay them down to count how many counters/beans are in each side, or part, of the bag.

Students record the whole and the parts using a number bond.

Then, they shake the bag and repeat to find a new combination.

Variation: This can be adapted to be a partner activity.

Player A shakes the bag, counts the beans on both sides and says, "8 is 5 and 3."

Player B writes the equation 8 = 5 + 3.

Players then switch roles and repeat the activity.

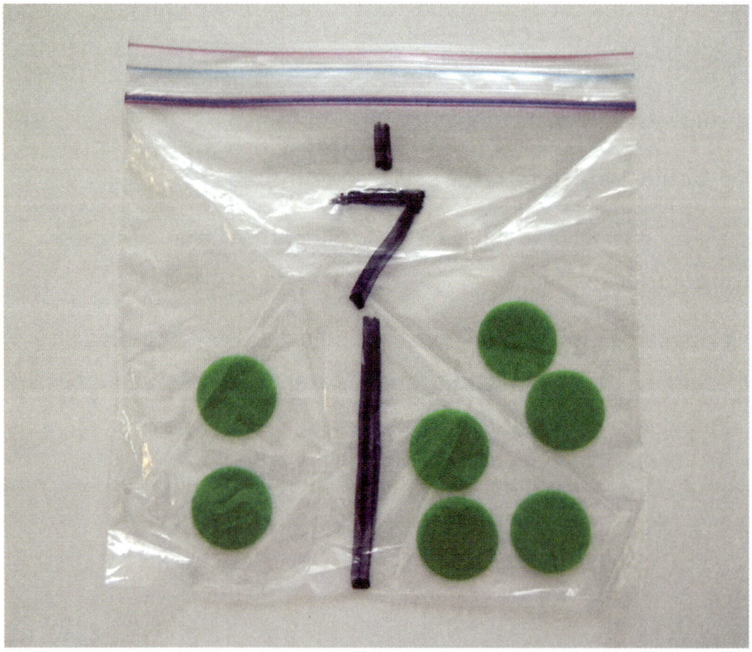

Here, 7 is shown as 2 and 5 in a bag. Students write the equation 7 = 2 + 5 to show the parts of the number.

Math Rack

A math rack is a tool that consists of two rows of beads and allows students to show numbers in various ways. Math racks can be used to show numbers 1–20.

Directions: Math racks are made using cardboard or craft foam, two pipe cleaners, and 20 small beads—10 each of two colors.

1. Cut a piece of craft foam to measure approximately 9 × 4 inches.
2. Use the end of a pipe cleaner to poke a hole near the top left of the foam and push the end through about two inches.
3. On the front, put 5 of the first color beads on the pipe cleaner and then 5 of the other color.
4. Poke another hole straight across near the top right and push the other end of the pipe cleaner through about two inches.
5. Repeat the process near the bottom to form another row of beads.
6. Start this row with 5 of the second color beads and then 5 of the first color.
7. Turn the rack over and twist the top and bottom ends of the pipe cleaners together.

Many students benefit from seeing the same concept modeled with different types of materials. Asking them to think about how the math racks are like part-part-whole mats or number bonds helps focus students on the big ideas that underlie all of these materials—that numbers can be broken apart in different ways.

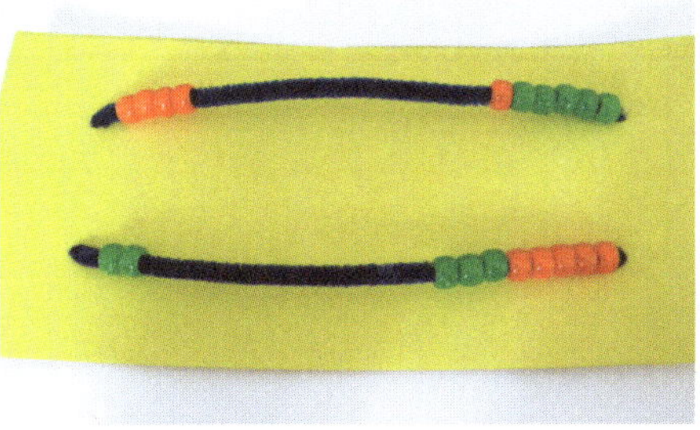

Beads are "parked" on the right side and are slid to the left to show a number. The front of the math rack shows 6 as 4 and 2. The back shows how the top and bottom pipe cleaners are twisted to secure the beads on the front.

Ask students to show you 8 on their math rack.

How did you show 8?

Did anyone make 8 a different way?

I see that Mike showed 8 as 6 beads on the top and 2 beads on the bottom.

Mark showed 8 as 2 beads on the top and 6 beads on the bottom.

How are these math racks the same?

How are these math racks different?

Students might make lists of different combinations for numbers.

Have students show their decompositions using equations like $8 = 6 + 2$ or $8 = 4 + 4$.

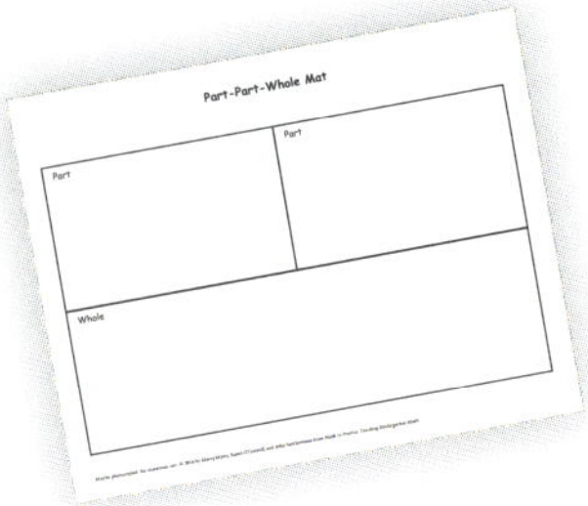

Drawing Parts on Mats

Part-part-whole mats are useful as a tool to organize the placement of manipulatives, but they can also be used as a tool for students to record parts and whole.

Have paper copies of the mats available for students to use.

These might be put in a plastic sheet protector so students can use a dry erase marker to draw pictures as they decompose numbers and then erase their drawings and begin again.

Part-part-whole mats do not need to be teacher-created. Students can create their own mats on a piece of blank paper.

Nontraditional Ten Frames

Nontraditional arrangements on ten frames provide students with other ways to see numbers. These combinations can be discussed as a class, labeled with numbers, and then written as a number bond or equation as students gain a better understanding of decomposing numbers.

Quickly flash ten frames with dots in nontraditional arrangements.

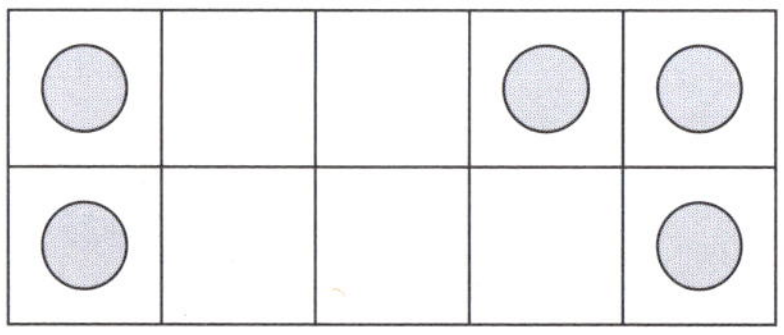

Ask:

What number did you see?
How did you see it?

Display the ten frame or draw it on the board. Have students come up and circle dots into groups to show how they saw the number. For example, 5 could be shown as 2 on the left and 3 on the right as in the example. Record the numbers as "2 and 3 are 5" or in an equation.

How Many Ways?

Use linking cubes to figure out how many ways there are to make 5 or 6 or 7 or another number.

Have students make a chain of 5 cubes and then break the chain into 2 parts, recording the numbers in each part.

$5 = 3 + 2$
$5 = 1 + 4$

How many ways can you do it?
Is 3 + 2 the same as 5? Explain.

Card Sort

Students select a number card and dot cards to equal the number.

Students select a 3–10 number card.

Students then make sets of dot cards that when combined equal the number on the card (e.g., the student selects the number 6 and then makes sets of dot cards that equal 6, like 2 and 4 or 1 and 5).

Students record their matches by drawing the two dot cards and labeling them with the number (the whole).

Then, they pick another number card and repeat.

Variation: This activity can be extended for students to write an equation that matches the dot cards.

Differentiation

This challenge activity gives practice to students who are ready to build equations to show their decompositions.

Domino Sort

Dominoes are a quick and easy tool for exploring parts (the dots on each side) and whole (the total number of dots) as students are challenged to tell the parts and whole and write the equations.

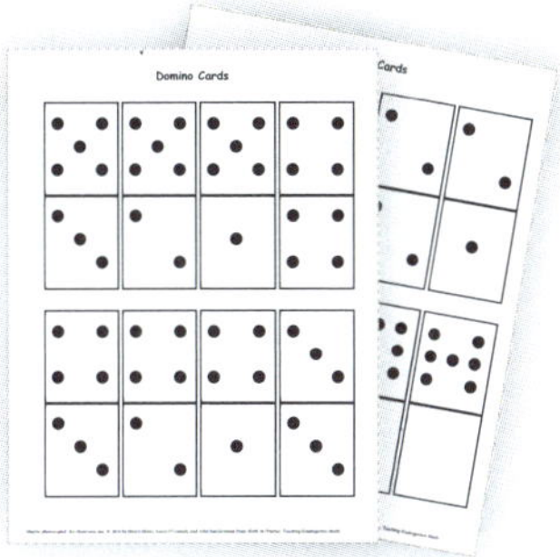

Give students 2–10 number cards (in a pile, facedown) and a set of dominoes (with pips, or dots, to a sum of 10). Have them spread the dominoes on the desk or table, faceup.

Version 1

Students work independently to sort the dominoes by placing the number cards on the table or floor and then sorting the dominoes by the total number of dots.

Students then record their work by writing equations to go with each number 2–10.

For versions 2 and 3, the number cards should be replaced in the pile after they are used.

Version 2

Player A chooses a number card and players work together to find a domino that has that number of dots (whole).

The players talk together to decide on the equation (e.g., 7 = 4 + 3).

Player A writes the equation.

Then Player B picks a card and they repeat the process, working together.

Version 3

Player A chooses a number card and looks for a domino to match his number.

If he selects one that equals 7, he says, "7 is 3 and 4," then records 7 = 3 + 4 on paper or a dry erase board.

Player B then takes a turn.

Play continues until time is up or all dominoes have been used.

Differentiation

As an extension, ask students why there are more dominos that equal 8, 9, and 10 than 2, 3, and 4.

FINDING THE NUMBER THAT MAKES 10

As students explore decomposing numbers, we give additional attention to decomposing 10 in order to build the foundation for many number concepts explored in first grade. Kindergarten students do not need to memorize the number pairs that equal 10. Instead, when given a single-digit number, they use concrete objects and drawings to find the number that adds to it to make 10.

Using Ten Frames to Make 10

Students use ten frames to visualize the number given and then find the number needed to make 10.

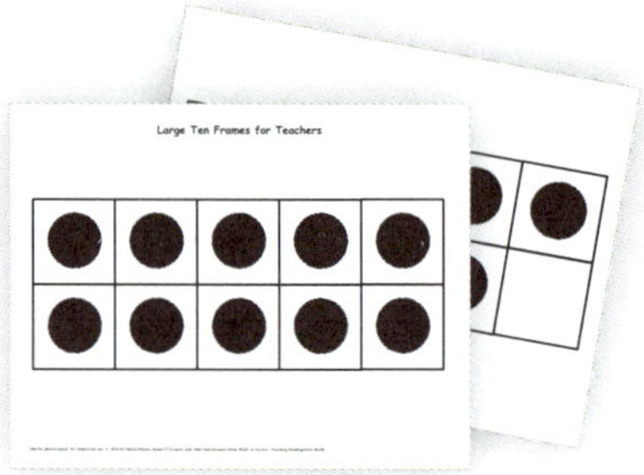

For this whole-group lesson, use a set of ten frames representing numbers 1–9 or prepare a slide for an interactive whiteboard that shows a ten frame with red and yellow circles or clip art that can be moved on to the frame as counters. If using paper ten frames, laminate them so they can be written on using dry erase markers.

Show the number 6 on the ten frame in a traditional arrangement (5 in one row, 1 in the next row).

What number do you see?

How do you know?

How many more circles do we need to show a full ten frame, or the number 10?

How do you know?

Draw circles on the laminated frame or add yellow circles on the whiteboard.

6 and 4 more is 10, so 6 and 4 together make 10.

How can we show 6 and 4 is 10?

Record a number bond to show 6 and 4 is 10.

Show the number 5 on the ten frame in a traditional arrangement.

What number do you see?

How do you know?

How many more circles do we need to show a full ten frame, or the number 10?

How do you know?

Draw circles on the laminated frame or add yellow circles on the whiteboard.

5 and 5 more is 10, so 5 and 5 together make 10.

Show me a number bond that shows what we did.

Observe students as they create number bonds to show that 5 and 5 is 10.

Can you find some other ways to make 10?

Use your ten frames and counters to find different ways to make 10.

Then show the numbers you found using a number bond.

Continue with other numbers.

Differentiation

To enrich this lesson, display numbers on the tens frames in nontraditional arrangements. Students work to identify the numbers shown and then name how many more are needed to make 10. For students who need additional support, use a ten frame for each student and 10 two-color counters. Students display the number shown using red counters and then fill the frame with yellow counters to show the number needed to make 10. The two colors of counters make the two parts very visible.

Revisiting these tasks over time (in whole or small group) gives students needed repetition.

Ways to Make 10

The Ways to Make 10 recording sheet provides students with an opportunity to explore the number combinations that equal 10.

Begin in a small group using 10 two-color counters and a ten frame for each student. Have each student place a red counter on the ten frame.

How many more do we need to make 10?

Complete your 10 frame using yellow counters.

After students place 9 yellow counters on the frame, give each student a Ways to Make 10 recording sheet. Model how to draw 1 red and 9 yellow counters. Then write the number sentence 1 + 9 = 10. Students will copy the drawing and write the numbers.

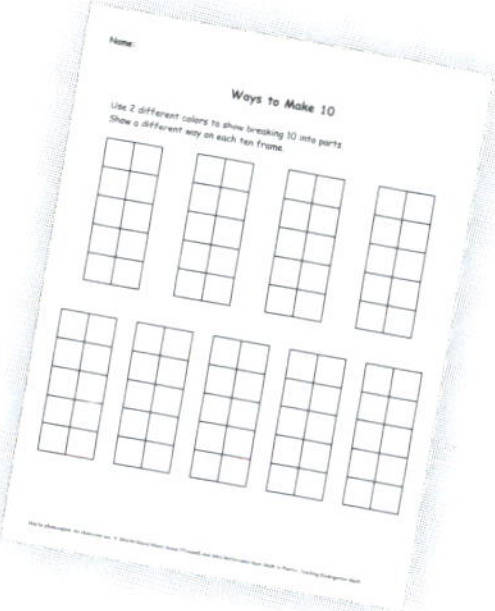

How many different ways can you find to make 10?

Find different ways to make 10 and record them on your sheet.

This student shows that 7 and 3 make 10 using two-color counters. The counters and ten frame allow her to visualize the parts of 10.

For additional formative assessment techniques, see Math in Practice: A Guide for Teachers, Chapter 5.

FORMATIVE ASSESSMENT

Show students a ten frame with 6 dots and ask,

How many more dots do I need to make 10?

The student may need to touch the ten frame to count the empty spaces, may pause to count the spaces without touching them, or may automatically say that 4 dots are needed to make 10.

Record observation notes of the student's response.

If the student is not able to determine the number needed on the frame to make 10, have the student work with 2-color counters on a ten frame to show numbers to make 10.

Exploring 10 with Bead Counters

In this activity, 10 beads (5 of two different colors) on a pipe cleaner provide hands-on practice breaking apart the number 10.

Students create a bead counter by stringing 5 beads of one color and then 5 beads of another color on a pipe cleaner and then pinching the ends of the pipe cleaner so the beads can move back and forth but won't fall off.

How many beads are on the pipe cleaner?
How do you know?

Push 4 beads to the side to make two groups of beads and ask students:

What do you see?

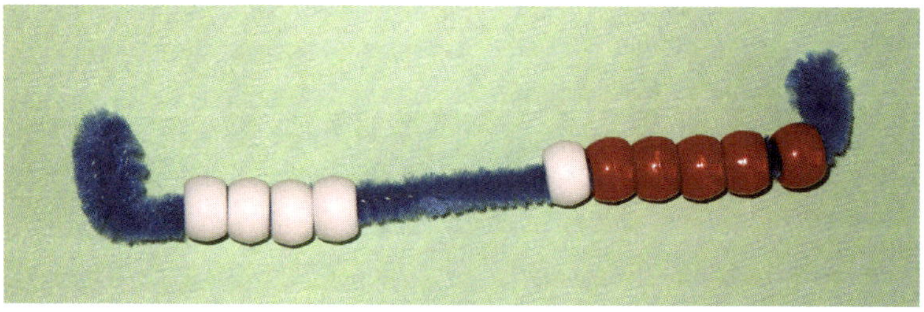

How do you know the second group is 6?
How many are on the other side? Tell me how you know.
What can you tell me about 10 when you look at my bead counter?
Are there still 10 even when we move the beads from side to side? Explain.

Record the equation to show the bead counter (10 = 4 + 6).

Have students use their bead counters to find other combinations for 10. Have them record equations to show what they find.

We are reviewing basic number concepts as we pose questions about how they know it is 6 or 4. By placing 5 beads of two different colors on the pipe cleaner, students are able to use their understanding of 5 to determine quantities just as they do when looking at ten frames.

Record their equations on the board.

What do you notice?

They may notice 6 + 4 and 4 + 6 or 2 + 8 and 8 + 2.

If you have 4 on one side, will you always have 6 on the other side? Explain.

Have students use their bead counters to test this idea with other number pairs like 3 and 7 or 2 and 8.

Does it always work?

Does that make sense?

Students may use examples to explain their thinking, like if you think of a ten frame and have 4 circles, you need 6 more to fill it; but if you have 6 circles, you need 4 more.

FORMATIVE ASSESSMENT

Give students ten frames that are partially completed and ask them to determine how many more counters would make 10.

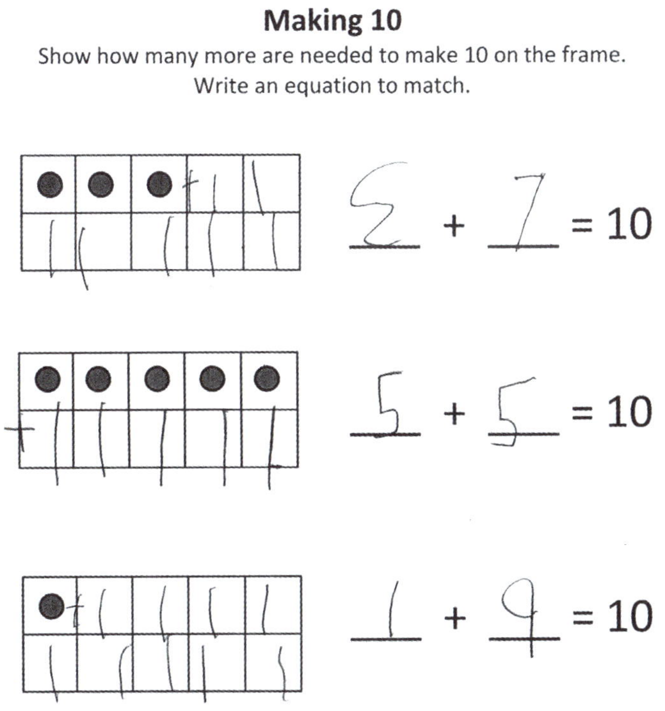

This student shows how many more are needed to make 10 and completes the equations. His strategy is to draw lines to show the number needed. Looking closely, you can see that he wrote a small plus sign between the dots and the lines to match the equation.

Additional Ideas for Support and Practice

The following ideas extend students' understanding of using objects and drawings to find the other number that makes 10 and provides meaningful practice.

Bead Counter Tens

Players use bead counters to find the parts that make 10.

Player A picks a 1–9 number card and then uses her bead counter to slide that number of beads to one side and find which number goes with that number to make 10 (e.g., she picks a 9 card and slides 9 beads to one side and sees that 1 bead is on the other side. She says 10 is 9 and 1).

Player B writes the equation 10 = 9 + 1.

Players then change roles and Player B picks a card and shows the number to Player A who writes the equation.

Players continue, alternating roles.

Stand in Tens

This movement activity can be used with the whole class or a small group to demonstrate the number needed to make 10.

In front of the whole class, have 7 students stand up. Say to the class:

There are 7 students standing.
How many more students do we need to make 10?
How do you know?
What number sentence could we write to show this?

Repeat with other numbers.

Making 10

Partners work together to make number combinations that equal 10.

Player A uses a ten frame mat in a plastic sheet protector and a dry erase marker and draws any number of dots on the ten frame (e.g., 6 dots).

Then Player A shows Player B the ten frame and asks, "How many more do I need to make 10?"

Player B figures out how many more are needed and says, "You need 4 more to make 10 because 6 and 4 is 10."

Then players switch roles.

Make-10 Concentration Partner Game

This game has students find combinations to make 10.

Prepare small ten frame cards for numbers 1–9 but with 2 fives.

Students place the cards facedown.

They take turns turning over the ten frame cards and looking for a pair of cards that equal 10.

When a pair is found, the student says the number combination and takes the cards.

Play continues until all cards are used.

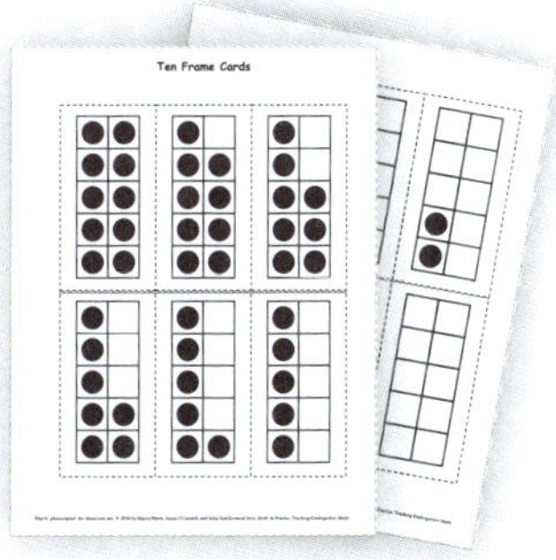

Make-10 Go Fish

Students play Make-10 Go Fish using two sets of ten frame cards for numbers 0–10. The ten frame cards have pictures to represent 0–10.

Each player gets five cards and places the rest of the playing cards in a "pond" between the players.

Players first look at their own cards to see if they have any pairs that equal 10. If they do, they place these pairs face up on the floor or table where they are working.

Then students take turns asking their partner for cards that they need to make 10. For example, if Player A is holding a 6, she will ask Player B, "Do you have a 4?"

If Player B has the card, he will give it to Player A.

If Player B does not have a 4, he will say "Go fish."

Player A will select a card out of the "pond."

The card selected from the pond can be combined with any cards they are holding to make 10.

Students continue the game taking turns asking for needed cards and "going fishing" until one of the players uses all of his or her cards to make sets of 10.

Talk About It/Write About It

Draw a ten frame with 3 counters on it. Show how many more are needed to make 10. Explain how you know.

Use objects or drawings to solve this problem.

Blake had 10 cupcakes. She puts them on 2 plates. How many could go on each plate? Explain how you solved it.

Differentiation
Setting the decomposition in a problem context provides a challenge for students who are ready for it.

Focus on 10

Use any of the activities from decomposing 2–9 to further explore decomposing 10.

- Shake and spill and fill a ten frame with yellow or red counters. Write equations.
- Look for dominoes that have 10 total dots and record the equations.
- Make chains of 10 linking cubes and break the chains to decompose 10.

Finger Flash Fun

In this enrichment task, students show the number on their fingers that makes 10 when shown a number card.

> **Differentiation**
> *At this level, students find the number that makes 10 using objects or drawings. This task focuses on students doing this mentally for those students who are ready to advance to this stage.*

Show the small group a numeral card (1–9) and have them show the number that, when added to the numeral card, equals 10 (e.g., if you show the group a 4, students would flash 6 fingers).

Say with the group:

4 and 6 are 10 or 4 + 6 = 10.

DECOMPOSING NUMBERS TO SOLVE PROBLEMS

Embedding decomposing numbers into story problems is perfect for applying the concept to real-life situations. Students might act out problems or draw pictures to solve them, using what they know about decomposing numbers. This type of problem is explored in greater detail in Module 8.

Turtles on a Log SHOW IT!

Students use work mats with logs and cutouts of turtles to model separating a group of turtles into two groups: turtles on the log and in the water.

For this whole-group lesson, prepare log mats and turtle cutouts (or counters to represent turtles) for each student.

Prepare a slide for an interactive whiteboard showing a log with turtles as clip art, or draw a log on the board and make a set of turtles to demonstrate on the board.

Begin by telling the students:

Do turtles move fast or slow?

We know that turtles move very slowly, but they are on the move all throughout the day.

Give each student a log work mat and 10 turtle counters.

> *Rather than demonstrating for students, we are asking them to show what they hear in the story.*

Let's see if you can show my story.

There are 5 turtles. Some were sunning on a log and others were swimming in the water.

How many were on the log and how many were swimming?

Have students show a possibility on their mat.

How many of your turtles were on the log? How many were swimming?
Do you still have 5 turtles?
Did someone do it a different way? Do you still have 5 turtles?
Is it okay to use different ways to show this?

Model student examples on the board.

Record different possibilities.

So, 2 turtles and 3 turtles is 5 turtles all together?
So, 4 turtles and 1 turtle is still 5 turtles?

Repeat with a different number of turtles (e.g., 8).

Let's try it with 8 turtles.
There are 8 turtles. Some turtles are in the water and some are on the log.
Use your turtles to show how many turtles are swimming in the water and how many are on the log.

Have students show a possibility on their mat, then ask them to describe their pictures, telling you the number of turtles in the water, the number on the log, and the number of turtles altogether.

Model student examples on the board.

Record different possibilities.

So, 5 turtles and 3 turtles is 8 turtles altogether?
So, 4 turtles and 4 turtles is 8 turtles?
Can we show the 8 turtles in different ways?

Continue with other numbers of turtles, allowing students to share their solutions and emphasizing that there are multiple ways to show each quantity of turtles.

To connect to and review previous skills, you might have students start with the turtles in the water and count on to find the total number of turtles.

Differentiation

Some students may need to work with numbers less than 5 and others will be ready to move to numbers less than 10.

Drawing Pictures to Represent Problems

Students draw pictures to show separating numbers.

For this small-group lesson, write problems ahead of time on chart paper or on slides for the interactive whiteboard. Students need paper and pencils, or individual dry erase boards and markers, to draw pictures to solve the problems.

Pose the problem:

There are 5 children.
Some are girls and some are boys.
How many are boys?
How many are girls?

How could we solve this problem?

Have students share their ideas:

Talk to your partner about a way you could draw a picture to show this problem and then draw one.

Observe as students work.

Allow students to use strategies that work for them. Some may draw 5 circles and then label 3 as girls. Others may draw 3 girls and then add 2 more to count up to 5.

If students need support getting started, ask:

What do you know about this problem?

What can you draw first?

What do you need to find out?

In this type of problem, students are trying to find both parts. There is more than 1 solution because the whole can be split in different ways.

After students have worked and have a solution, ask:

What were we trying to find out in this problem?

Turn and talk to your neighbor. Tell him or her how you solved the problem.

Can someone show us your drawing and explain what you did?

Select a couple of students who drew pictures and solved the problem in different ways to share. Have them show their drawing and tell how they did it. Ask questions to clarify, like these:

So Dakira drew 5 circles, and then she labeled 3 with Gs. What do the Gs represent?

Then she labeled the rest with Bs. What do the Bs represent?

So how did she know how many boys there were?

Turn and tell a partner what number sentence you could write to match this picture.

Point to the number sentence as you ask more questions:

What is this 5?

What is the 3?

What is the 2?

Point to $5 = 3 + 2$ as you connect students' findings back to the equation:

There are 5 students; 3 are girls and 2 are boys, so $5 = 3 + 2$.

Did anyone get a different number of boys and girls? Does it have to be 3 girls and 2 boys?

Have students with a different solution share and discuss their method and solution.

Rita drew 5 circles and she circled some. Tell us what you did, Rita.

Turn and tell a partner what number sentence you could write to match this picture.

What is this 5?

What is the 4?

What is the 1?

Does it make sense that there is more than one answer? Why?

What does this remind you of? (Activities breaking apart numbers.)

Make connections to breaking apart towers or separating beads or spilling red/yellow counters.

For additional practice have students draw pictures to solve:

Molly has 6 juice boxes.

Some are apple juice and some are orange juice.

How many of each?

FORMATIVE ASSESSMENT

Give students the Popsicle Problem to show ways to decompose 7.

> Grandma has 7 popsicles.
>
> Some are orange and some are green.
>
> How many of each color are there?

Students can decompose 7 with any combination and represent the problem with manipulatives or by drawing a picture.

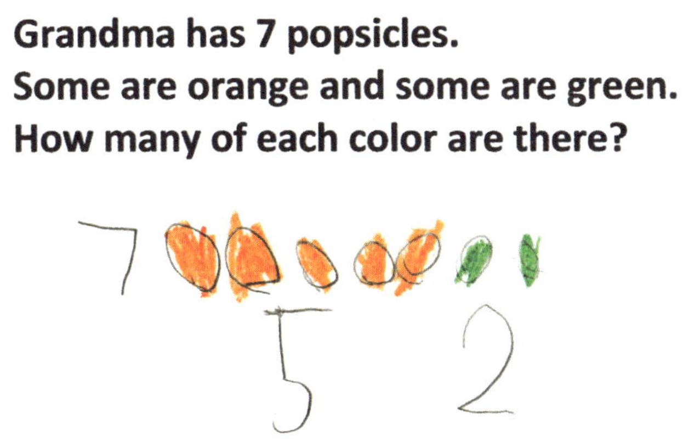

This student solves the problem showing 7 as 5 and 2. She colors 5 orange circles and 2 green circles. She labeled 7 as the whole and the parts as 5 and 2. Although one possibility was all that was required, this student could be challenged to show another possible answer to the problem.

This student uses connecting cubes to visualize and solve the Popsicle Problem. He shows multiple solutions including 2 orange and 5 green, 6 green and 1 orange, and 4 green and 3 orange. He now needs to draw a picture to represent his solutions.

Vocabulary

break apart
combine
decompose
equation
join
number bond
number sentence
part
part-part-whole
same as
separate
whole

ONLINE RESOURCES

General Resources

Additional Problems
Decomposing Numbers Student Observation Notes
Domino Cards
Dot Cards 0–10
Large Ten Frames for Teachers
Number Cards 0–10
Part-Part-Whole Mat
Ten Frame Cards
Ten Frame Mats

Resources for Specific Activities

Dot Card Sums Recording Sheet
Drawing a Picture to Decompose Recording Sheet
Log Mat and Turtle Cutouts
More Making 10 Student Sheet
Popsicle Problem Sheet
Ways to Make 10 Recording Sheet

These resources are available at http://hein.pub/MathinPractice, keycode MIPGK.

Additional Ideas for Support and Practice

The following ideas extend students' understanding of using decomposition to solve problems and provide meaningful practice.

Problem Partners

Have students work with partners to solve problems using objects or drawings.

Materials: tool basket with cubes, counters, ten frame, individual dry erase board, and marker for each pair

There are some dogs.
There are some cats.
There are 10 pets.
How many of each?

There are 10 dogs.
Some are big and some are small.
How many of each?

Stuart washes the dishes after dinner.
He washes 7 dishes.
Some of them are plates and some of them are cups.
How many plates and how many cups did he wash?

Lizzy has 2 favorite colors, green and pink.
All of her dresses are green or pink.
If she has 9 dresses, then how many of each color could she have?

Zach and Debra are decorating cookies.
They put chocolate chips on some and sprinkles on the others.
They make 10 cookies.
How many have chocolate chips and how many have sprinkles?

MODULE
7

Understanding Addition

About the Math K.OA.1; K.OA.2

Students are introduced to the operations of addition and subtraction in kindergarten. Math operations are active. Exploring the meanings of the operations in active ways helps our students realize this.

The key ideas focused on in this module include:

- **understanding the operation of addition as *putting together* or *adding to***
- **showing addition situations by acting out problems, using objects to model problems, or drawing pictures of problem situations**
- **recognizing and writing addition equations to represent problem situations.**

In kindergarten, we explore the operations of addition and subtraction by acting them out, representing them with objects, drawing pictures to show the actions, and, finally, showing them with numbers and symbols. Understanding addition and subtraction is not about key words, but rather about key ideas. When students understand the actions of the operations, they know when to add or subtract.

There are two situations, or story structures, that show the operation of addition: *putting together* or *adding to*. In both cases, we are looking for a total, or a sum. By posing lots of situations that show these actions, modeling the actions, and discussing what they have in common, our students begin to understand the action of addition. At this level, we focus on showing ways to find the sum, or total, of two quantities. Students are able to use concrete materials and drawings to find the total. To simplify

Although sum is the official term for the answer when we add, the language can be confusing in kindergarten because of the already-known word some, which essentially means a part of something. Using the term total is fine at this level and may help to alleviate the confusion with the term sum.

the situations, we limit the numbers to a total of five in early problems, and eventually a total of ten as students gain experience and confidence. In this way, our focus remains on building understanding and strategies to find the total.

ADDITION AS ADDING TO

Acting out addition problems is a great way to get students up and moving and allows them to experience the action of addition. One way we think about addition is *adding to* a quantity. Problems that show addition as *adding to* might include the following:

4 students are wiggling. 1 more wiggles. How many are wiggling?

3 students are in line to get a drink of water. 2 more students get in line. How many are in line?

2 students are jumping. 2 more students start to jump. How many are jumping?

Using concrete materials clarifies the concept of *adding to*. Using objects, including fingers, allows our students to see the action of addition.

There are 4 crayons in the basket. Put 2 more crayons in the basket. How many are in the basket?

There are 3 cubes in a train. Put 1 more cube on the train. How many cubes are there in the train?

Show me 3 fingers. Show me 1 more finger. How many fingers?

Drawing pictures of problem situations also helps students interpret the meaning of *adding to*.

Draw 3 balloons. Draw 1 more. How many balloons?

Draw 2 hearts. Draw 2 more. How many hearts?

As students explore the addition structure of *adding to*, we might record the expression (3 + 2) or the equation (3 + 2 = 5), allowing them to see the process in this symbolic way. In early experiences, we record the expressions because our focus is on developing an understanding of the operation. Later, with our support, students explore this new abstract way to show addition. Although their first experiences are active and visual, they will later link the numbers and symbols to these active experiences.

ADDITION AS PUTTING TOGETHER OR JOINING

Joining, or *putting together*, is another action that shows addition. This addition structure is about composition, putting two quantities together to get a total. Again, acting out scenarios helps students visualize this addition structure.

3 boys are standing. 2 girls are standing. How many children are standing?

2 children are holding reading books. 2 children are holding math books. How many children are holding books?

Using concrete materials helps students understand what it means to *join* or *put together*.

There are 4 red crayons. There are 2 blue crayons. How many crayons?

There are 3 blue cubes. There is 1 green cube. How many cubes?

Drawing pictures to show *putting together* allows students to clarify the action.

Draw 2 little dogs. Draw 2 big dogs. How many dogs?

Draw 2 pink hearts. Draw 2 red hearts. How many hearts?

As students explore the addition structure of *putting together*, we record the expressions and talk about the action that indicates addition.

Through story problems, students see what it means to add. Through retelling and visualizing the problem situations, they begin to see the actions that tell them when addition makes sense. As students move together to act out a situation, or as crayons are added to other crayons in a basket, they visualize the addition process. And supporting their visualizations through pictures, part-part-whole mats, and number bonds (see Figure 7.1) provides visual images to clarify the process. We help students transition to representing addition in an abstract way as we first record expressions and equations and then ask students to write them, introducing them to symbols, like + and =, that allow them to show addition situations using numbers and symbols.

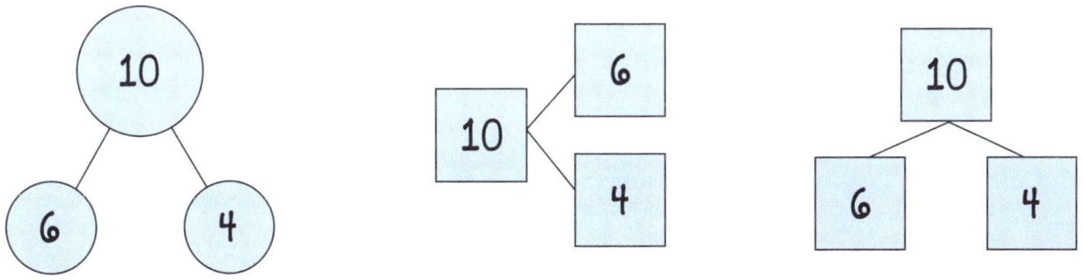

Figure 7.1 Number bonds show the two addends and the sum. They can be shown in different arrangements but always show the connection between the parts and the total.

Exploring the Progression

PREVIOUS		NOW		NEXT
Preschool Possible experience with the concept of addition or the meaning of addition equations	»	**Grade K** Understanding that addition has two situations: *putting together* and *adding to*; recognizing or writing equations to show addition	»	**Grade 1** Expanding understanding of addition by exploring problems with unknowns in various positions

Learning Goals

I can solve addition problems by acting out problems or using objects.

I can solve addition problems by drawing pictures.

I can write equations to show addition problems.

Teacher Tips for Problem Solving: Addition

As we work with word problems, we use a variety of instructional strategies that help students comprehend the problems before they begin to solve them. When students comprehend word problems, they understand the situation, as well as the known and unknown information. With this understanding in mind, students are able to choose an operation to solve the problem, rather than randomly adding or subtracting the numbers they see in the problem. Following are some instructional strategies that help primary students comprehend and solve word problems effectively.

MULTIPLE READS FOR A VARIETY OF PURPOSES

In kindergarten, you will most likely be reading problems to your students, but we try to keep the text short to allow our emergent readers to begin to read the problems on their own. Reading the problem several times shows your students that you reread problems for different purposes. The first time you read the problem to students, you might simply ask them to tell you what it is about, ignoring the specific data (the numbers). Do they know that things are being put together or being added to? You may read it again to see how the numbers fit in and ask students to act it out or make a model to show the specific data. After students tell you a solution, you may read it again and ask them if their answer makes sense.

First read: What is happening?

Another read: How do the numbers fit in? How can I show it?

Read it again: Does my answer make sense?

STUDENTS ACTING OUT PROBLEMS

Kindergarten students love to act out stories. Having them work together to act out a word problem is a powerful way to build comprehension. Students show us their understanding of the problem situation as they act it out.

USING TECHNOLOGY TO ACT OUT A PROBLEM

Interactive whiteboard technology allows us to import photos of students or fun clip art to use in our problems and has the added benefit of our being able to show movement as we move the graphics to show addition situations. This technology engages students, allows them to visualize the story problem, and provides a tool for them to actively show what is happening in the problem.

MODELING A PROBLEM WITH COUNTERS

Using counters is more abstract than acting out problems together. Using counters to represent the problem data, arranging or moving them to show what is happening in a problem, is an important step for students before transitioning to drawings.

RETELLING THE EVENTS OF THE STORY

Kindergarten students begin to recognize the sequence of events in stories by thinking about what happens at the beginning, in the middle, and at the end. This strategy can also be used with word problems, especially when there are events and actions involved in the word problem (e.g., *adding to* and *putting together*).

What is happening in the problem?

Can you retell it in your own words?

How does it start?

Then what happens?

Three Ways

A three ways graphic organizer helps students make connections between word problems and related models and equations. In this graphic organizer (see Figure 7.2), the word problem is provided and students add a model and an equation to match it.

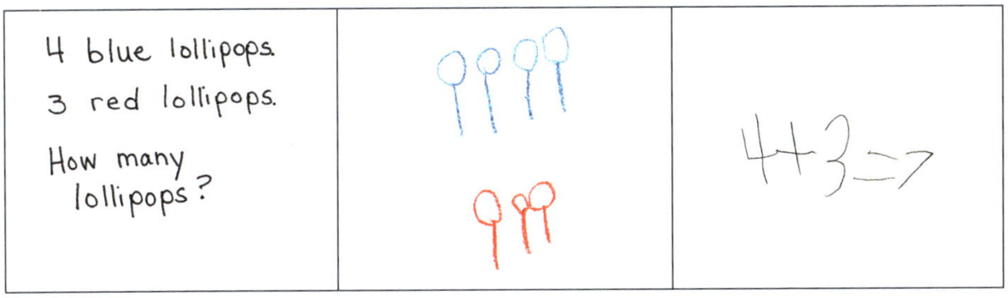

Figure 7.2 This student draws a model to show the problem and writes the equation to match it.

Part-Part-Whole

Using part-part-whole graphic organizers as work mats gives students a way to model and organize information from the word problem. Students retell the problem by looking at their part-part-whole model. (See Figure 7.3.)

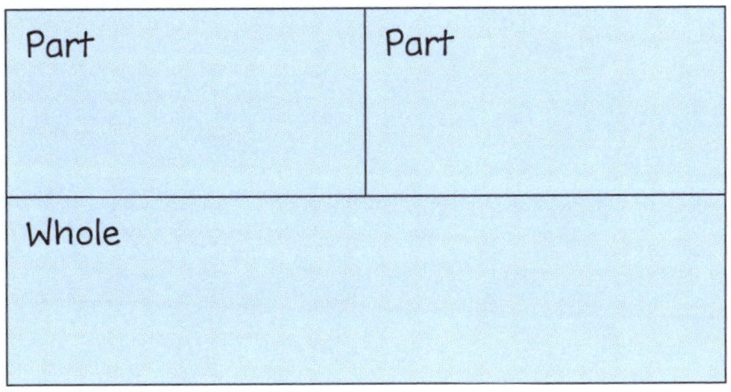

Figure 7.3 Students place counters or draw pictures in sections of the mat and are able to retell what is happening in the story.

Lessons in This Module

Acting Out or Using Manipulatives to Show *Adding To*

Act Out *Adding To*

Adding To with Bears

Acting Out or Using Manipulatives to Show *Putting Together*

Act Out *Putting Together*

Putting Together Using Manipulatives

Part-Part-Whole Mats with Objects to Show *Putting Together*

Using Manipulatives and Part-Part-Whole Mats

Drawing to Show *Adding To* and *Putting Together*

Drawing to Show Addition Problems

Using Symbols for Addition

Using Number Bonds to Show Addition

Representing Addition with Equations

DRAWINGS AND DIAGRAMS

Helping students relate models to drawings and diagrams is an important step in the development of a problem solver.

Simple Drawings

Students begin by drawing pictures of the balloons, ice-cream cones, or birds in the word problem. The simple act of drawing the problem data allows students to visualize what is happening in the problem. Once they are able to see the data, they are better able to find the solution.

Representational Drawing

Students may begin by drawing circles to represent the balloons, ice-cream cones, or birds in the problem, understanding that each circle represents one of the items and recognizing that they can visualize the problem without making detailed pictures.

Part-Part-Whole Mats or Number Bonds

Students can draw simple part-part-whole models to represent the problem situation. Number bonds are similar to the diagrams in that they show the parts and the whole, but they have numbers in each section rather than drawings (see Figure 7.4). Labeling these diagrams helps keep students focused on the problem data.

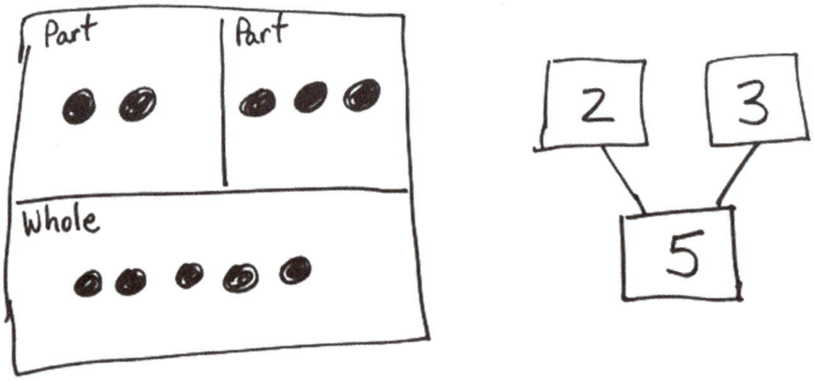

Figure 7.4 Number bonds allow students to see the part–whole relationship in a more abstract way.

Ideas for Instruction and Assessment

ACTING OUT OR USING MANIPULATIVES TO SHOW *ADDING TO*

Students begin to learn about the addition structure of *adding to* by acting out the problem or using concrete manipulatives to solve it. The learning is active and ongoing to provide students with numerous opportunities to understand the operation.

Act Out Adding To

Students learn about *adding to* by acting out problems in this whole-class lesson.

Prepare problems according to the number of students in your kindergarten class, ensuring that all students have an opportunity to participate actively in an acting out problem.

Before the lesson, write problems on chart paper or display them on the whiteboard.

Today we are going to listen to some math stories and solve them.

3 girls were sitting at the reading table.

Then 2 more girls sat at the reading table.

How many girls are at the reading table now?

How can we act this story out?

How many girls were at the reading table first?

Heather, Jessica, and Maria, go over and sit at the reading table.

There are 3 girls at the reading table.

How many more girls go to the table?

Then 2 more girls sat at the reading table. Janae and Abby, go over to the table.

How many girls are at the reading table now?

Let's count: 1, 2, 3, 4, 5.

Stand near each girl as you count.

It can be difficult for kindergarten students to sustain focus as others are acting out problems. To increase engagement, students who are not acting can use their fingers to show the numbers as you count.

There are 5 girls at the reading table.

Heather, Jessica, and Maria are 3 girls. Janae and Abby are 2 girls.

3 girls and 2 girls are 5 girls.

Let's try another story.

4 boys were lined up at the door.

5 more boys line up at the door.

How many boys are lined up at the door?

How can we act this out?

How many boys need to line up first?

4 boys go stand in line at the door. There are 4 boys.

Then how many more boys line up?

5 more boys go stand in line at the door. There are 5 more boys.

How many boys are lined up at the door?

Stand near each boy as you count.

Let's count: 1, 2, 3, 4, 5, 6, 7, 8, 9. There are 9 boys lined up at the door.

4 boys and 5 boys are 9 boys.

Bears can be substituted with beads, cubes, or other manipulatives.

The focus of this lesson is acting out the problems to solve, not reading or writing equations.

For additional strategies to build students' problem-solving skills, see Math in Practice: A Guide for Teachers, Chapters 1 and 2.

Adding To with Bears

Students use concrete materials to solve *adding to* problems.

For each student in the small group, have bags of manipulatives, such as bears, insects, or beads. Students use the manipulatives to act out the addition stories. It is helpful to write the problems on the board or chart paper for the students to refer to as they work.

Here's a story about some bears.

Brittany has 3 bears and then she got 4 more bears.

How many bears does Brittany have now?

Turn and tell your partner what is happening in this story.

How could you use bear counters to show the story?

Show me how you can solve this problem.

Reread the story.

How many bears does Brittany have?

Show me her 3 bears.

What happened next?

Show me the 4 more bears.

What are we trying to find out?

Reread the problem.

What are we trying to find out?

How do we know that? (Look at them. *Or,* Count them.)

So let's count all of the bears: 1, 2, 3, 4, 5, 6, 7. She has 7 bears.

3 and 4 more is 7.

Reread the problem.

Did we solve the problem?

Pose the problem:

Carla has 4 bears.

Then she got 2 more bears.

How many bears does she have now?

Tell your partner what this story is about.

Show me how you can use counters to show the story.

Reread the story.

How many bears does Carla have?

How do you know?

What did you do first? (I showed 4 bears.)

Then what did you do? (I put 2 more.)

Then what did you do? (I counted them all to see how many.)

Let's count all of the bears: 1, 2, 3, 4, 5, 6. She has 6 bears.

4 and 2 makes 6.

Reread the problem.

Did we solve the problem?

Let's try one more story.

Paolo has 7 bears and then he got 3 more bears.

How many bears does Paolo have now?

Tell your partner the story.

How can you use the bear counters to show the story?

Reread the story.

How many bears does Paolo have now?

Turn and talk with your partner. Tell each other how you solved the problem.

Who can tell us how you and your partner solved the problem?

Differentiation

Numbers can be +1 or +2 for students who need to practice with smaller numbers to understand the operation. For enrichment, start the problem with the smaller number and add the larger number.

FORMATIVE ASSESSMENT

Pose a problem for students to solve with a set of manipulatives other than teddy bears.

Record observational notes of student work and responses.

Joe has 4 insects.

Then he finds 3 more.

How many insects does Joe have now?

Students should be able to show 4 insects and 3 insects and count them as 7.

If a student struggles to add the 4 and 3, try simpler addends like 4 and 1.

Before asking for an answer, have the student show you what the problem looks like (the 4 insects and 1 insect) and have him tell you what he needs to find out (how many insects altogether). Ask him how he can find that out (count them). Watch as he counts to be sure he has accurate counting skills.

Use the observational notes page.

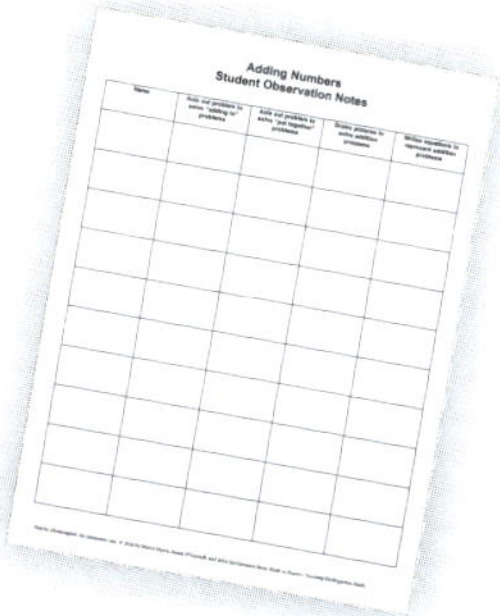

Additional Ideas for Support and Practice

The following ideas extend students' understanding of using manipulatives to solve *adding to* problems and provide meaningful practice.

Mouse Count

Students add 1 connecting cube as 1 more mouse is added to a jar.

Each student needs a small pile of connecting cubes to help illustrate the story.

Before Reading:

Ask the students:

What does it mean to add?

During Reading:

Read *Mouse Count* by Ellen Stoll Walsh.

As you read each page, have the students add "1 more mouse" to follow the story.

For example, the book shows 5 mice. Then 1 more mouse is put in the jar.

Students will show 5 cubes, and then connect 1 more to make the total of 6 mice. Discuss that 5 and 1 more makes 6.

Pause when 10 mice are in the jar. Talk about adding 1 more mouse. There are now 10 mice in the jar.

Continue reading the story.

After Reading:

Ask the students:

What happened when we added 1 more mouse? (The number of mice increased by 1.)

Adding To with Connecting Cubes

Students roll a number cube to practice *adding to* to make trains of 10.

Students need a set of connecting cubes and a 1–6 number cube.

All students begin with 1 connecting cube.

Students roll and take that number of cubes (e.g., roll a 3 and take 3 cubes).

The number rolled is then *added to* the original cube and students verbalize the addition that is happening (e.g., 1 cube and 3 more makes 4 cubes).

Play continues until students make a train of 10 cubes.

Then, they set that train aside and begin again with 1 cube to make another train of 10.

Students try to make as many trains as possible in the time given.

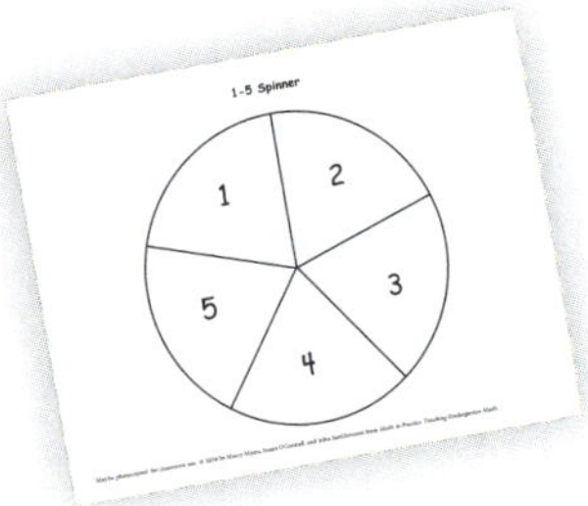

Spin and Add Game

Partners take turns spinning numbers to add to a ten frame.

Each pair needs a spinner with numbers 1–5, a blank ten frame, and counters.

Students play with a partner.

Students take turns spinning the spinner and putting manipulatives on the ten frame.

During each turn, the students must say how many counters they have on their ten frame, how many they are *adding to*, and how many they have total (e.g., "I have 4 bears, I am adding 2 more bears, now I have 6 bears.")

Students continue playing until one partner has filled his ten frame.

Then, they clear their frames and start again.

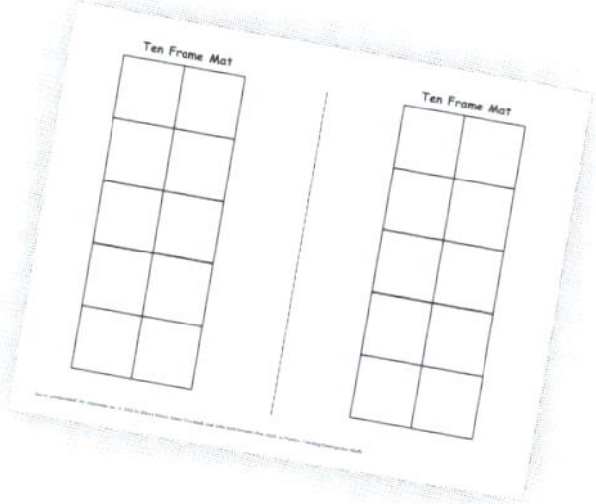

ACTING OUT OR USING MANIPULATIVES TO SHOW PUTTING TOGETHER

Putting together problems combine two groups to make a new group, rather than *adding to* an existing group. Again, the learning is active and ongoing to provide students with numerous opportunities to understand the operation.

Act Out Putting Together

Students will learn about *putting together* by acting out problems in this whole-class lesson.

Before the lesson, write problems on chart paper or display them on the whiteboard.

The focus of this lesson is acting out the problems to solve, not reading or writing equations.

We have some math stories to listen to and solve with our class.

2 students are jumping on the floor.

4 students are jumping on the rug.

How many students are jumping?

How can we solve this problem? (We can act it out.)

What should we do first?

Okay, here are 2 students jumping on the floor.

What do we need next?

Let's have 4 students jump on the rug.

How many students are jumping?

Move near the students as you count to find the total.

We have 1, 2 students (on the floor) *and 3, 4, 5, 6 students* (on the rug).

How many students are jumping?

We put 2 students together with 4 students to get a group of 6 students jumping.

Here's another story to try.

6 students are coloring with markers.

3 students are coloring with crayons.

How many students are coloring?

Give 6 students a marker to hold and 3 students a crayon to hold.

How can we solve this problem?

What should we act out first?

Have the 6 students pretend to color with markers.

Now what do we need to act out?

Have the 3 students pretend to color with markers.

What should we do to find how many students are coloring?

Okay, count with me.

Stand near the students as the class counts.

1, 2, 3, 4, 5, 6, 7, 8, 9. How many students are coloring?

We put 3 students together with 6 students to get 9 students.

Continue with other problems that model *putting together*.

As students become familiar with acting out the problems, you can begin to write equations to represent the stories.

Thinking Through a Lesson

Putting Together *Using Manipulatives*

Students use manipulatives to explore the *putting together* addition structure.

For each student in the small group, have bags of manipulatives such as insects, bears, or beads.

Young children are often tempted to play with manipulatives instead of using them to explore the math. To increase students' focus, set a timer and give them one minute of "exploration time" with the manipulatives. Explain that once they hear the timer ring, students will be expected to use the manipulatives to show the math stories.

Today we are going to solve some math problems.

Instead of acting them out, we are going to find another way to show the stories and find the answers.

Write the problems to match the type of manipulatives students will be using.

Read the problem:

Jazir has 5 red beetles.
He has 3 yellow beetles.
How many beetles does he have?

Incorporate the names of your students in the problems you pose. Their engagement in the activity immediately increases.

SMP1 *Tell your partner what this story is about.* (Jazir has red and yellow beetles.)

Give students some insect counters.

Can you use these counters to show me Jazir's beetles? SHOW IT!

Retelling the story to partners is a first step in comprehending the story problem and gets all students immediately involved in the lesson.

Reread the story. Watch as students show the beetles using the counters.

Because students at this level have limited reading skills, rereading the problem at different points throughout the lesson is helpful, particularly when they are asked to listen for different purposes (e.g., to find the data, to decide what they are trying to find out, to see if the answer makes sense).

Turn and show your partner your 5 red beetles.

Show your partner your 3 yellow beetles.

We showed what Jazir has.

SMP4

Now, listen to the story to see what we need to find out.

Reread the problem.

Tell your partner what we are trying to find out. (How many beetles he has.)

Are the red ones beetles? (Yes.)

Are the yellow ones beetles? (Yes.)

So we need to find out how many we have with all of the beetles? (Yes.)

How could we do that? (Count them.)

Tell your partner how many beetles Jazir has.

Watch as students count and tell their partners the answer.

How many beetles does Jazir have? (8.)

How did you know that? (We counted.)

Did you count just the red ones? Did you count just the yellow ones? (We counted all of them.)

Through questions and actions, we are trying to help students see that putting together groups is adding.

Show how you push together the 5 beetles and the 3 beetles and then count all 8.

Students do not need to write their own equations at this point, but recording them begins to give them familiarity with the format.

Write $5 + 3 = 8$ on the board.

Point to the numbers in the equation as you say:

You added 5 beetles and 3 beetles to get 8 beetles.

SMP2

Through our talk, we help students see what the numbers in the equation represent.

Reread the story.

Did we find out how many beetles he has? (Yes, he has 8.)

Here's another story to solve.

Jordan has 7 butterflies and 3 grasshoppers.

How many insects does she have?

Tell your partner what is happening in this story. (Jordan has butterflies and grasshoppers.)

Listen again to see what we are trying to find out.

Reread the story.

What are we trying to figure out? (How many insects she has.)

Are butterflies insects? (Yes.)

Are grasshoppers insects? (Yes.)

How can you use your insect counters to show the problem?

Turn and show your partner.

Where are the 7 butterflies?

Where are the 3 grasshoppers?

How will you know how many insects she has? (Count them all.)

SMP3 SMP6

Why do you count them all? (They are all insects.)

So, we need to know the total number of insects? (Yes.)

We help students develop the language of addition by using words like total as we talk about the story problem.

Have students count and tell their partners how many insects Jordan has. Ask the group:

How many insects does Jordan have? (10.)

How do you know that? (We counted.)

What did you count? (The butterflies and grasshoppers.)

Why did you count all of them? (We wanted to know how many she had with both. *Or,* We wanted to know the total.)

Do not tell them to count them all. Watch and see what they do. Then discuss it afterward.

Praise students who repeat the language of addition at appropriate times during the discussions.

Again, push two sets of counters together to show the adding motion.

Write $7 + 3 = 10$ on the board.

Point to the numbers in the equation as you say:

You added 7 butterflies and 3 grasshoppers to get 10 insects.

Reread the story.

Did we find out how many insects she had? (Yes, she had 10.)

If time permits, continue with another problem:

Dylan has 2 red grasshoppers

He has 5 green grasshoppers.

How many grasshoppers does he have?

Differentiation

You might have some students show and solve this problem with partners while you focus attention on a few students who need extra support to show the problem with their counters or count to find the total.

Have students retell the problem, show it with the counters, and explain how they found the answer.

Remind students they have been *putting together* the two groups to find the total.

Remind them that they have been adding.

FORMATIVE ASSESSMENT

Pose the problem:

I was cleaning up the counters and I found 2 red counters and 4 yellow counters.

How many counters did I find?

Give students counters and ask them to show you what is happening in the problem.

Record observational notes of student work and responses.

Students should be able to show 2 red and 4 yellow counters and count them all as 6. They should be able to tell you that they needed to count all of them to find the answer.

If the student has trouble remembering 2 and 4, write *2 red* and *4 yellow* on the board or paper for the student to refer to as she works.

Use the observational notes page.

Additional Ideas for Support and Practice

The following ideas extend students' understanding of using manipulatives to solve *putting together* problems and provide meaningful practice.

Domino Addition Game

Students add using dominoes and place the domino in the cup that shows the total.

Students need 10 cups labeled with numbers 1–10 and dominoes whose dots/pips total 1–10.

Students can play this game with a partner or independently.

The student picks a domino and adds the dots/pips together to find the total.

Then, they place the domino into the cup labeled with the total.

Play continues until all dominoes are placed into the correct cups.

Ten Frame Addition

Partners work together to add on a ten frame to find the greatest (or least) total.

Each pair will need a blank ten frame, 2 dice, and 2 different colors of manipulatives.

In a small group, students work with a partner.

Before playing, the group decides if teams are playing for the greatest or lowest total.

Then, partners roll, place manipulatives on the ten frame, and decide on their total.

Each partner rolls a die and adds that many manipulatives to the ten frame (e.g., Player A rolls a 4 and Player B rolls a 5).

Player A puts 4 (red) manipulatives on the ten frame and Player B puts 5 (blue) manipulatives on the ten frame.

The pair says that their ten frame shows 4 and 5 put together makes 9.

Whichever pair in the small group has a number that best meets the original goal (greater or fewer) gets a point or sticker (e.g., team 1 has a total of 7 and team 2 has a total of 9. If the goal was "greater," then team 2 would get the point or sticker for that round).

Problem-Solving Journals

Students solve *put together* problems using ten frames and manipulatives and then write about the problems in their journals.

Provide each student with manipulatives and ten frames when solving the story problems.

Jen and Molly are playing with dolls.
Jen has 4 dolls and Molly has 3 dolls.
How many dolls do the girls have?

Solve by using manipulatives on the ten frame and then record your answer with a picture.

Travis collects seashells on the beach.
On Monday he collects 5 shells and on Tuesday he collects 3 shells.
How many shells did Travis collect?

Use the cubes to show your answer and then record it in your problem-solving notebook.

Derik is helping his mom plant flowers.
They plant 7 flowers in the front yard and 3 flowers in the backyard.
How many flowers did they plant?

Use the cubes to show your answer and then record it in your problem-solving notebook.

Part-part-whole mats are not used to represent adding to problems because that problem structure adds on to the original group.

PART-PART-WHOLE MATS WITH OBJECTS TO SHOW PUTTING TOGETHER

Students can use part-part-whole mats to solve *putting together* problems. The format of the mats models *putting together* by joining two parts to make the whole.

Paper plates divided into three sections also work well for part-part-whole activities.

Using Manipulatives and Part-Part-Whole Mats

Students solve problems using manipulatives and part-part-whole mats.

Each student in the small group needs a part-part-whole mat and cubes.

Write the problems on chart paper or on the board for students to refer to as they work.

Read the problem:

I have 4 red apples and 3 green apples.
How many apples do I have?

Tell your partner what this problem is about.

Let's show the problem with cubes.

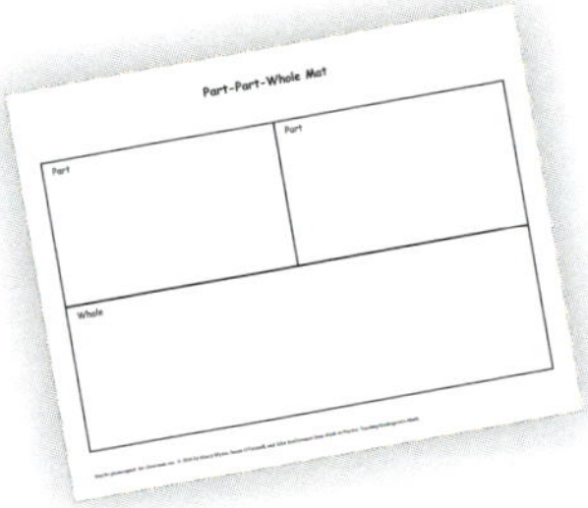

Give students a part-part-whole mat and some cubes.

How could we use this mat to show the problem?

I'll read the problem again and then you can show me the story on your mat.

Reread the problem.

Use your counters to show me the red and green apples.

Where will you put the cubes to show the 4 red apples?

Where will you put the cubes to show the 3 green apples?

How can we show the total number of apples?

How can we find out how many apples I have altogether?

1, 2, 3, 4, 5, 6, 7. I have 7 apples.

Did we solve the problem?

Reread the problem.

I put together 4 and 3 and got 7.

7 what? (Apples.)

Does that sound right?

Did the part-part-whole mat help us? How?

Pose the problem:

There are 5 baseballs and 2 basketballs.

How many balls are there?

Tell your partner what this problem is about.

Solve this problem using your cubes and part-part-whole mat.

Allow students to work independently to solve the problem.

How did you use your cubes and part-part-whole mat to solve the problem?

You had a part that was 5 and a part that was 2.

You added 5 and 2 to equal 7.

Reread the problem.

Did we solve it? How many balls are there?

Here's another problem to solve.

There are 3 girls on the soccer team.

There are 4 boys on the soccer team.

How many children are on the team?

Retell the problem to your partner.

How can you use a part-part-whole mat to solve it?

Students should begin to understand how to put each part of the problem on the mat and combine the parts to make the whole.

Observe how students count and place cubes in part of the mat.

Observe if students are able to combine the parts to make the whole and count to find the total.

Have students share solutions.

Students should be familiar with part-part-whole mats from using them to decompose numbers in Module 6.

These guiding questions help focus students' thinking as they comprehend the problem and show it on the mat.

Observe students as they are working with the cubes. Make sure they are using the correct numbers and placing them correctly on the mat.

As students are working, emphasize that they are working to find the total number of apples—the whole group of apples. Part of that group is red, and the other part is green

As students work independently, ask individual students to explain how they are solving the problem. Students with full understanding will explain that they are placing 5 baseballs on part of the mat and 2 baseballs on the other side of the mat. Then they are combining the baseballs to make a whole group of 7. If students lack understanding, prompt them as necessary.

Differentiation

Some students might need support to put the cubes in the parts of the mat or need prompting to combine them to find the whole.

Problems using part-part-whole mats can be ongoing throughout the year. Eventually students will begin to write equations representing the mats.

Reread the problem.

Did we solve it? So, how many children are on the team?

How does using a part-part-whole mat help you solve the problem?

This student uses a part-part-whole mat to solve 3 + 4. He puts 3 green chips in one part and 4 blue chips in the other part. Then he moves the chips to the whole section to find the answer.

FORMATIVE ASSESSMENT

Give students a part-part-whole mat with cubes.

Show how you can solve this problem using the mat and cubes.

I have 3 chocolate cookies and 6 sugar cookies.

How many cookies do I have?

Repeat the problem as necessary for students or write "3 chocolate" and "6 sugar" on the board for students to refer to as they work.

Record observational notes of student work and responses.

Students will put 3 cubes in one part and 6 cubes in the other part. They will combine the parts in the whole section and count 9 cubes.

For students who have trouble adding 3 + 6, use smaller numbers such as 3 + 1, 2 + 2, or 1 + 4 to assess their use of a part-part-whole mat.

Use the observational notes page.

Additional Ideas for Support and Practice

The following ideas extend students' understanding of using manipulatives and part-part-whole mats to solve *putting together* problems and provide meaningful practice.

Part-Part-Whole Activity

Students use number cards and manipulatives on a part-part-whole mat to find the total.

Materials: part-part-whole mat, manipulatives, number cards 1–5 (two sets, ten cards total) for each student or pair

The student picks two number cards at random and places them in the "part-part" sections of the mat. The student will build both numbers with manipulatives.

Then the student moves the manipulatives to the "whole" section of the mat to find the total (leaving the number cards in the "part-part" sections).

The student says what he has shown.

For example, the cards 1 and 3 were picked.

The student would say, "1 and 3 make 4 or 1 + 3 = 4."

If playing with a partner, each student will pick only one card and build one set in the "part-part" section.

Then both partners will total the "whole" and tell what was shown.

Sample *Putting Together* Problems

Students solve problems using two colors of counters on a part-part-whole mat. Each student needs a part-part-whole mat and two colors of counters.

Pose problems for students to solve.

Bill makes 5 cookies.
Jane makes 4.
How many cookies did they make?

Samantha catches 3 frogs.
Jane catches 2 frogs.
How many frogs did they catch?

Tom is looking for bugs.
He sees 4 bugs in the garden and 6 bugs on the sidewalk.
How many bugs did Tom see?

DRAWING TO SHOW ADDING TO AND PUTTING TOGETHER

When students have had practice using concrete manipulatives to act out *adding to* and *putting together* problems, the next step is drawing pictures. Students begin to draw simple pictures to illustrate the story problems. Students benefit from ongoing practice in using pictures to represent addition.

Drawing to Show Addition Problems

The pictures do not have to be detailed because the focus of a math picture is to show what is happening in the problem. The pictures can be as simple as circles or squares to represent the problems.

Students will solve *adding to* problems by drawing simple math pictures to represent the problem.

For this whole-class lesson, write problems ahead of time on chart paper and allow space for drawing pictures to solve.

Pose the problem:

Rachel has 3 flowers.
Then she picks 2 more flowers.
How many flowers does Rachel have now?

Retell this story to your partner.
How could we show this problem when we don't have manipulatives?
Could we show this problem with pictures?
Draw a picture to show the problem.

Have students share their pictures.

Draw 3 flowers and then 2 flowers on the chart paper.

Does the picture help us know how many flowers Rachel picked?
How many flowers did she pick?
How do you know?
What if I drew this . . . ?

Draw circles to represent the flowers, near the original flowers.

Does this picture help us know how many flowers Rachel picked?
How are the two pictures similar?
How are they different?
What do you think my circles are supposed to be?
Math pictures don't have to be fancy or detailed. They just have to show what is happening in the problem.
So, how many flowers did Rachel pick? How do you know?
Let's count the circles: 1, 2, 3, 4, 5.

Reread the problem.

Did we solve it? How many flowers did Rachel pick?

Rachel has 5 flowers. 3 and 2 equals 5.

Can someone remind us what we did to solve the problem?

How would I know how many flowers she picked from looking at your picture?

Could you add anything so I wouldn't have to count them, so I would just know how many she picked?

You could write 5 to tell me how many flowers she picked.

Here's another problem that we can solve by drawing a picture.

Oscar has 6 cars.

Then he got 4 more cars.

How many cars does Oscar have now?

Have students retell the problem to partners.

What do we need to find out?

What could we draw to represent the cars?

Do we need to draw pictures that look just like cars? Why?

Could we use circles or squares or rectangles to represent the cars?

Reread the problem.

What should I draw first?

Then he got 4 more cars.

What should I draw next?

How many cars does Oscar have now? Could someone come up and count the cars for us?

Jenna says that there are 10 cars. Is she correct? How do you know?

Reread the problem.

Did we solve it?

Can I write something to let people know how many cars there are?

Write the number to show how many cars Oscar has.

Note: This lesson focused on *adding to* problems. Follow a similar format to focus on problems with *putting together* situations like the following:

The clown has 4 red balloons and 4 blue balloons.

How many balloons does the clown have?

Luke has 7 pieces of candy.

Jack has 2 pieces of candy.

How many pieces of candy do the boys have?

Allow students time to draw the pictures to represent the problems and talk about how they used the pictures to find their solutions.

At times you might suggest a way to represent items, but students should be allowed to represent the problem data in ways that make sense to them. It may take some students longer to give up detailed drawing and move to more abstract representations.

As students progress with drawing pictures, they can write the numbers shown. For example, they would write 6 and 4 for the cars and 10 as the total. That will be an important foundation for writing equations to represent the problem.

Drawing pictures to solve problems should be ongoing. Problem solving can occur as a whole class or in small groups so students become proficient in drawing simple pictures to solve story problems.

FORMATIVE ASSESSMENT

Pose the problem:

> The clown has 2 yellow balloons and 5 green balloons.
> How many balloons does the clown have?

Drawing a picture helps this student solve a *putting together* problem. The student draws 7 balloons and labels them to show the total. She also writes a number sentence. Although she shows a reversal for the number 5, her understanding of the number is evident.

Additional Ideas for Support and Practice

The following ideas extend students' understanding of drawing pictures to solve *adding to* and *putting together* problems and provide meaningful practice.

Shaving Cream Pictures

Students solve problems by drawing pictures in shaving cream.

Materials: inexpensive cookie trays, shaving cream, dish towels or paper towels

> Each student gets a cookie sheet with a blob of shaving cream.
>
> They are instructed to use one hand (dominant hand) to spread out the shaving cream on the tray. (The other hand is for holding the tray steady on the table.)

Students will listen to a story problem and solve by drawing a quick picture in the shaving cream (shapes, dots, dashes, etc.).

After discussing the answer and the thinking used to solve the problem, students can spread the shaving cream back out and be ready for a new problem.

Ten Frame Progression

Students learn to represent items without the gridlines of a ten frame.

The ability to organize items or drawings helps students count them correctly. When using ten frames, the items or drawings are organized by the frame itself. Slowly weaning students off the ten frames and talking about how they can organize drawings or materials helps them learn to become more organized in their representations.

Initially, when solving addition problems, give students a ten frame template to use as a guide for organizing items or counters (see Figure A).

Once students become skilled at using this tool to organize the objects, remove the vertical lines of the ten frame (see Figure B).

Finally, remove all inner lines of the ten frame and just provide a rectangle space for students to organize their materials (see Figure C).

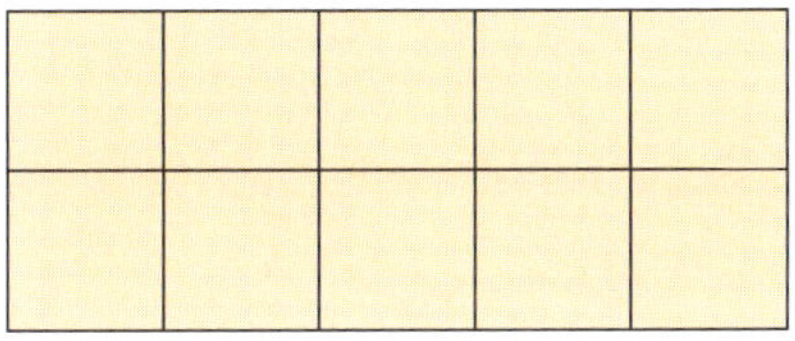

Figure A

Figure B

Figure C

Talk about how they place items within the rectangle and how rows allow them to more accurately count what is there.

This progression allows students to develop an understanding of the organization needed when displaying items on their own.

Students will then be able to use an open space and create organized drawings or collections of objects when solving problems.

USING SYMBOLS FOR ADDITION

Students begin to explore number representations that show problem situations and make connections between representing problems with concrete objects and pictures and using these number representations.

Using Number Bonds to Show Addition

Students explore number bonds to show problem data.

This whole-group lesson can be presented using an interactive whiteboard or chart paper. Students can use dry erase boards with markers to draw pictures.

Pose the problem:

The dog has 6 bones.
Then he finds 4 more bones.
How many bones does the dog have now?

Retell the problem to your partner.
How many does the dog have?
How could we use your dry erase boards to solve this problem?
Draw pictures to solve the problem.
Okay, show me your boards. What did you draw?

Observe as the students work to draw accurate pictures. Encourage them to label the 6 and 4 bones and write the total.

Have several students share their drawings and explain what they represent.

I see the 6 bones and the 4 more bones.
How many bones does the dog have? How do you know?
I see that many of you wrote 10 on your boards. The dog has 10 bones. 6 and 4 are 10.

Praise students who labeled their boards with the numbers to go along with their pictures.

Draw a number bond on the board.

Choose one of these number bond diagrams. It could be best to start with the total on top to connect to part-part-whole mats.

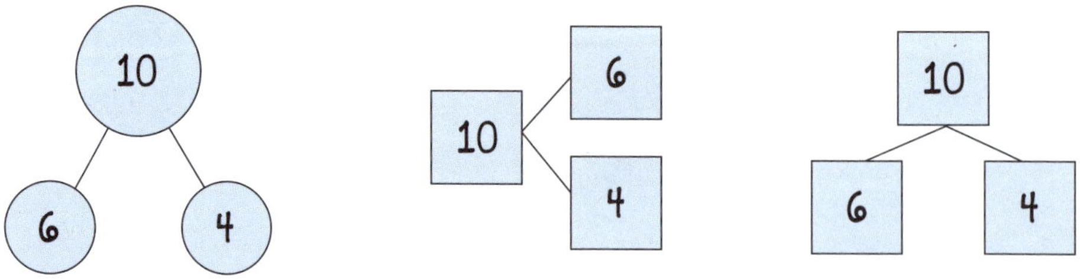

What do you think this shows? How might it show what is happening in our problem?
This is called a number bond.

How does the number bond represent the problem?

What tells you that 10 is the total?

Try to draw a number bond on your dry erase board to show 6 and 4 are 10.

Show me your number bonds.

Did we solve the problem?

Reread the problem.

How many bones does the dog have?

Great! Erase your boards. Let's try another story.

Lindsey has 5 crayons.

Betsy has 4 crayons.

How many crayons do the girls have?

Tell your partner what this problem is about.

What picture could you draw to show how many crayons the girls have?

Remember to label each girl's crayons with the number they have and write the total.

Show me your boards.

Have students share and explain their work.

Did you show Lindsey's 5 crayons and Betsy's 4 crayons?

Did you include numbers with your pictures?

I see that 5 and 4 are 9.

Did we solve the problem?

Reread the problem.

How many crayons did they have?

How can we use a number bond to show it? Draw one on your board.

Turn and talk with your partner.

Show your neighbor the number bond and say, "5 and 4 are 9."

Draw a number bond on the board to show 5 and 4 are 9.

Does our number bond match the problem? What are the 5 and 4? What is the 9?

Differentiation

Some students may need to demonstrate the problem using connecting cubes with the number bonds. The cubes can be placed as the "parts" and then moved to the "whole" to show the part-part-whole connection. Check that students are accurate with the number bond.

Representing Addition with Equations

When students have had numerous experiences with manipulatives, drawing pictures, and using number bonds to understand addition and solve problems, they are ready to write equations to represent the problems.

Students work with manipulatives or draw a picture and show a number bond. This lesson can occur in small or large groups.

Pose the problem:

There are 4 birds in the tree, and 2 more birds land in the tree. How many birds are in the tree?

What is this problem about?

What do we need to find out?

Tell your partner a way you could do that.

Show me how you can solve it.

Observe students as they work.

Ask them to also use a number bond to show the problem.

Turn and talk to your neighbor. Explain to your neighbor how you solved the problem.

What does your number bond show?

Draw a number bond on the board to show the problem.

You have seen me write number sentences when we solve problems.

Write a number sentence by the number bond.

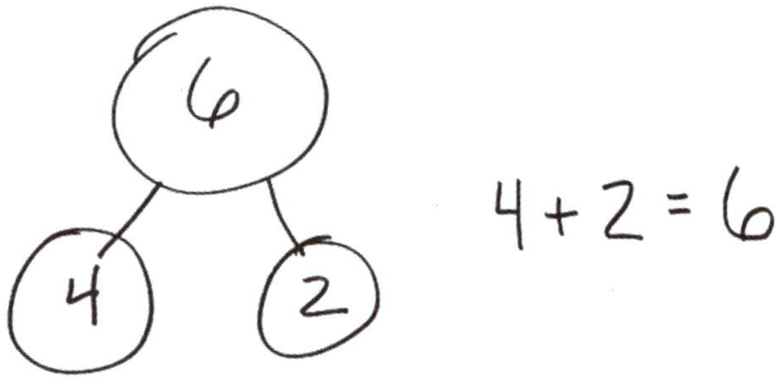

How is a number sentence like a number bond?

What do you notice about the information from the problem, the 4 and 2?

How do I show that 4 and 2 make 6 in a number bond?

Point to the 4, 2, and 6 on the number bond as you say that 4 and 2 make 6.

How do I show that 4 and 2 make 6 in a number sentence?

Point to the 4, 2, and 6 on the number sentence as you say that 4 and 2 make 6.

We write sentences with words to tell stories, and we can write sentences with numbers to show math stories. These number sentences are also called equations.

Write *number sentence* and *equation* on your word wall or Math Talk chart. Record *4 + 2 = 6* next to the words.

Point to the plus sign and say:

This is a plus sign.

The plus sign tells us that we are adding. We added 4 plus 2.

Plus *means adding.*

Add the word and symbol for *plus* to the Math Talk chart.

Point to the equal sign.

This is an equal sign.

The equal sign tells us that it equals the total.

It shows that 4 birds plus 4 birds equals 8 birds.

4 birds plus 4 birds are the same number of birds as 8 birds.

Add the word and symbol for *equal* to the Math Talk chart.

Write 4 + 2 = 6 by your problem.

Did we solve the problem?

Reread the problem.

How many birds are in the tree?

Nice job! Let's try to solve another problem.

Addie has 3 books.

Alexa has 3 books.

How many books do the girls have?

Tell your partner what the problem is about.

Solve the problem with a picture, then do a number bond to show the total.

Allow students time to work. Have them share their answers and show their pictures and number bonds.

Draw the number bond on the board.

Let's write a number sentence to show that.

Write 3 + 3 = 6 on the board by the number bond.

This says 3 plus 3 equals 6. That means 3 and 3 are 6.

Write the number sentence by your picture.

Did we solve the problem?

Reread the problem.

How many books did the girls have?

As we continue adding to *and* putting together, *we will write more number sentences.*

Observe as students write the equation. This is an introduction to equations. It will take some time for students to be able to compose them on their own. Continue to emphasize the meaning of the plus sign and equal sign, explaining equal as being "the same as."

This type of lesson is repeated as students continue solving problems throughout the year. It is important in kindergarten that students act out, use manipulatives, and draw pictures to solve problems. As the year goes on, they begin to connect number representations (number bonds and equations) to those concrete and pictorial representations.

FORMATIVE ASSESSMENT

Give students a dry erase board and marker.

Have them solve the following problem.

> There are 2 bananas and 8 oranges.
> How many pieces of fruit are there?

Record observation notes of student work.

Use the observational notes page.

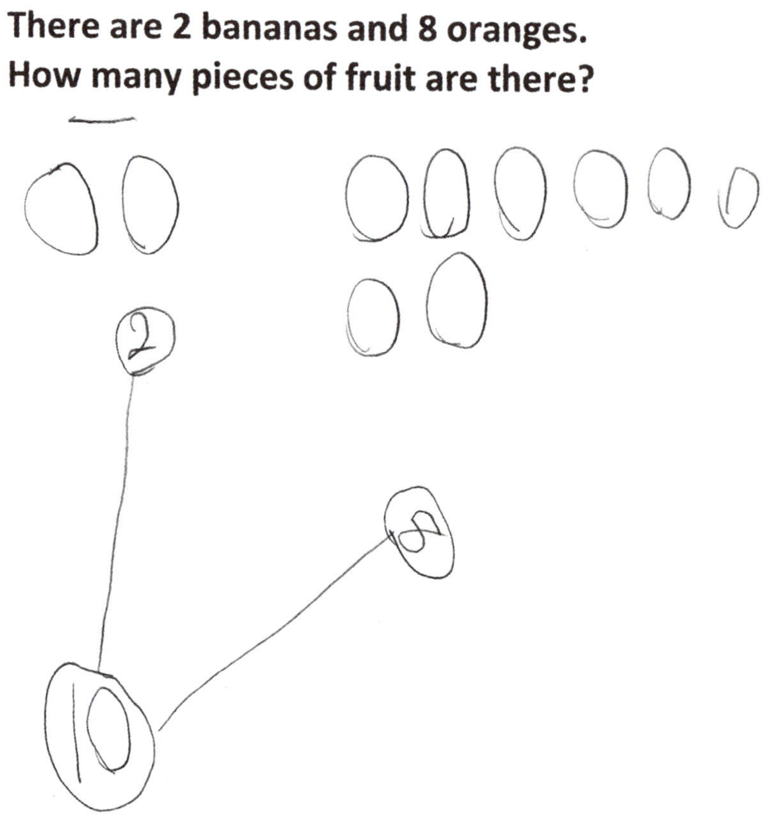

This student draws a picture to show 2 bananas and 8 oranges and is able to say that there are 10 pieces of fruit. He also draws the number bond, showing the 2 and 8 as parts in smaller circles and the 10 as the whole in a larger circle.

**There are 2 bananas and 8 oranges.
How many pieces of fruit are there?**

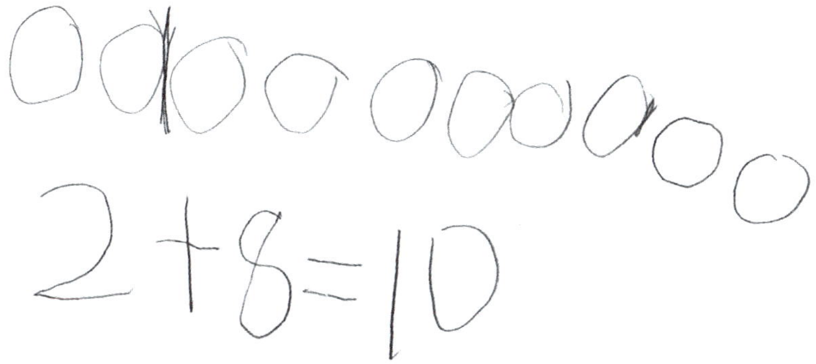

This student draws circles to represent the 2 bananas and 8 oranges and separates them by drawing a line between the groups. He correctly writes a number sentence to match his drawing and gets the correct solution of 10 pieces of fruit.

Additional Ideas for Support and Practice

The following ideas extend students' understanding of number bonds and equations to solve addition problems and provide meaningful practice.

Writing Equations to Match Pictures

Students write equations to match pictures drawn by the teacher.

Each student needs a dry erase board and marker.

On the board or chart paper, draw 3 circles and 4 squares.

Students write the equation that is shown in the picture. For example, students will write 3 + 4 = 7.

Ask students:

Why did you write 3 + 4?
How do you know the total is 7?

Continue with other pictures and number combinations.

Some students may benefit from practice with simpler +1 pictures.

Talk About It/Write About It

What does it mean to add?

Draw a picture to show 3 + 1.

There are 6 frogs in the pond. What addition problem could we write about the frogs? Draw a picture to show the addition problem.

There are 4 dogs at the park. What addition problem could we write about the dogs? Draw a picture to show the addition problem.

Eliminate It

Students work with a partner to decide which of the following does not belong with the others, then use math reasoning, numbers, or models to convince their classmates that their thinking makes sense.

4 + 0	2 + 2
	1 + 4

Fill in the Blanks Addition

Differentiation

For students who are beginning to understand number sentences, this activity gives them practice completing them.

Students select number cards to make number sentences and then solve.

Draw an outline for an addition equation on a large index card or piece of paper (laminated).

You might use either of these templates:

___ + ___ = ___

or

___ = ___ + ___

Have students pick two number cards (1–5) and place them on the lines on either side of the plus sign (e.g., 5 + 3 = ___).

Have students use objects or drawings to solve the problem and then write the correct answer with a dry erase marker on the line.

5 + 3 = 8

or

8 = 5 + 3

Vocabulary

add	*plus*
add to	*put together*
equal	*same as*
equation	*total (preferable to sum at this level)*
number bond	
number sentence	

ONLINE RESOURCES

General Resources

Adding Numbers Student Observation Notes

Additional Problems

Part-Part-Whole Mat

Ten Frame Mats

Resources for Specific Activities

1–5 Spinner

Balloon Problem Sheet

Fruit Problem Sheet

These resources are available at http://hein.pub/MathinPractice, keycode MIPGK.

MODULE

8

Understanding Subtraction

About the Math K.OA.1; K.OA.2

Addition and subtraction do not appear in a separate module in this book to indicate that subtraction should be taught separately from addition, but simply to allow more space for activities and problems related to addition and subtraction and to allow us to share additional insights for teaching these operations. Addition and subtraction are integrally connected. They show inverse actions. Addition is *putting together*; subtraction is *taking apart*. Addition is *adding to*; subtraction is *taking from*.

The key ideas focused on in this module include:

- **understanding the operation of subtraction as *taking from* or *separating***
- **showing subtraction situations by acting out problems, using objects to model problems, or drawing pictures of problem situations**
- **recognizing and writing subtraction equations to represent problem situations.**

Just like with addition, students' introduction to subtraction should be through varied activities that show the action of subtraction, including acting out scenarios, using objects, and drawing pictures to show the subtraction process. Through stories and problems, students experience the action of subtraction. They see subtraction in context.

In kindergarten, students explore two subtraction structures: *taking from* and *taking apart*. In *taking from* situations, students are given a total and see that a quantity is *taken from* that total. They are determining what remains. By acting out problems, students see varied possibilities for *taking from*.

5 students are jumping. 1 stops jumping. How many are still jumping?

4 girls are standing. 2 sit down. How many are still standing?

Using concrete materials allows students to explore *taking from*. Students might use cubes or counters to show these situations:

Molly had 3 cookies. She ate 1. How many did she have left?

Bailey had 5 pennies in her hand. She dropped 2 of them. How many were still in her hand?

And drawing the total and then crossing out what is removed allows students to explore this structure.

There were 5 balloons. 2 popped. How many were left?

4 dogs were in the yard. 1 ran away. How many were left?

As students remove objects from a group, lower fingers that were raised, or cross out dogs that were drawn on their paper (see Figure 8.1), they are visualizing *taking from* situations.

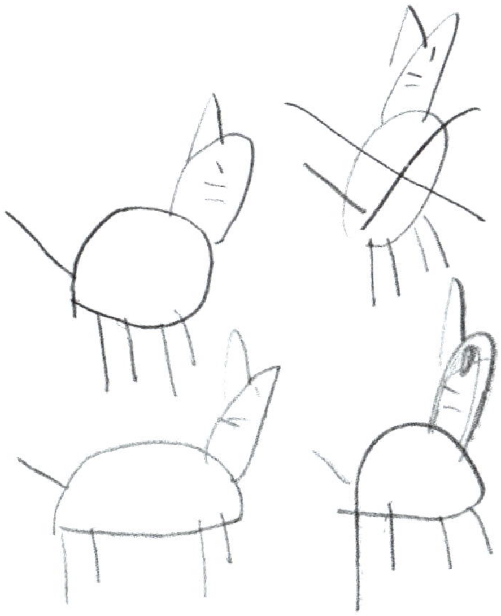

Figure 8.1 A student draws 4 dogs and then crosses one out to show that 1 dog ran away.

Another subtraction structure for kindergarten students is *taking apart*, or *separating*. At this level students are simply decomposing a quantity to find its parts, which are both unknown.

There are 4 balloons. Some are red and some are blue. How many of each?

There are 5 children. Some are boys and some are girls. How many of each?

There are 3 dogs. Some are little and some are big. How many of each?

At this introductory level, taking apart feels more like addition than subtraction. We even use an addition equation to show the two unknown parts. As students move to first grade, however, the whole and one part are known and students use the known information to find the missing part (e.g., There are 4 balloons. 3 are red and the rest are blue. How many are blue? 4 − 3 = 1). Taking apart problems are presented in this module because the structure is linked to subtraction in first grade.

In *taking apart* problems, there is more than one possible answer. This structure connects to their experiences decomposing numbers (breaking them apart) to find that 5 can be broken into 3 and 2 or 4 and 1. Exploring these problems by acting out situations or using concrete materials, and then having students record their findings through drawing or numbers, allows them to find the multiple answers to these problems. Recording the equation as $5 = 3 + 2$ makes sense for *taking apart* (*separating*) problems because we know the total and then solve the problem to find the two parts that are unknown (the addends). And writing equations, such as $5 = 3 + 2$, helps our students understand the meaning of the equal sign as *same as* and know, right from the start, that equations can be written in different ways. (See Figure 8.2.)

Figure 8.2 Acting out problems allows students to see what is happening in the problem. This group acts out a *separating* subtraction problem about 3 students. Two students have folded arms and 1 student has his arms at his side.

As with addition, this is our students' first exposure to expressions $(5 - 2)$ and equations $(5 - 2 = 3)$. Initially, we record the equations for our students to allow them to see this new notation, still placing our main focus on understanding the subtraction story. Having class discussions focused on the symbols used in subtraction (–, =) and what they mean, along with frequently asking students to tell us what the symbols mean, builds their understanding and confidence. As their comfort with the symbols increases, we encourage them to write the equations to represent subtraction stories.

Exploring the Progression

PREVIOUS		NOW		NEXT
Preschool		**Grade K**		**Grade 1**
Probably no experience with the concept of subtraction	»	Exploring the operation of subtraction by acting it out, representing with objects, and drawing; recognizing or writing equations to show subtraction problems	»	Expanding understanding of subtraction by exploring comparison problems (How many more? How many less?) and take apart/separate problems

Learning Goals

I can solve subtraction problems by acting out the problems or using objects.

I can solve subtraction problems by drawing pictures.

Teacher Tips for Problem Solving: Subtraction

Subtraction word problems can be a challenge for kindergarten students. To help them understand subtraction, we continue to use instructional strategies that help them learn to comprehend word problems before they solve them. When students comprehend word problems, they understand the situation, as well as the known and unknown information. With this understanding in mind, students can skillfully choose the operation to solve for what is unknown. The following instructional strategies, identified in Module 7 to support students' understanding of addition word problems, help students comprehend and solve subtraction word problems as well.

MULTIPLE READS FOR A VARIETY OF PURPOSES

Skillful problem solvers read word problems more than once. They may read the first time, ignoring the numbers, to get a general sense of what is happening in the problem, such as whether something is being *taken away* or *put together*. They may read it again to see how the numbers fit in and then make a model to represent it. After solving, they may go back and read it again to make sure their answer makes sense. For kindergarten students, this means that we reread problems for them to allow them to hear the problems at different points during their process of solving them.

First read: What is happening?
Another read: How do the numbers fit in? How can I show it?
Read it again: Does my answer make sense?

STUDENTS ACTING OUT PROBLEMS

Having students work together to act out a word problem, especially when it is about them, is a powerful way to build comprehension and helps them see the action in the story.

Lessons in This Module

USING TECHNOLOGY TO ACT OUT A PROBLEM

Interactive whiteboard technology engages students, allows them to visualize the story problem, and provides a tool for them to actively show what is happening in the problem as they move graphics on the board.

MODELING A PROBLEM WITH COUNTERS

Using counters to represent the problem data, arranging or moving the counters to show what is happening in a problem, is an important step for students before transitioning to drawings.

RETELLING THE STORY EVENTS

Kindergarten students are often asked to retell the stories they read, frequently chunking information as the beginning, middle, and end of the story. This strategy can also be used with word problems, especially when there are events and actions involved in the word problems (as in *take from*, *put together*, *take apart*, or *add to* structures).

What is happening in the problem?
Can you retell it in your own words?
How did this story begin?
What happened next?
What are we trying to find out?

Retelling the problem helps students identify the information they know and figure out what they don't know.

DRAWINGS AND DIAGRAMS

Simple drawings of objects, representational drawings using circles or Xs to represent objects, and graphic organizers, like part-part-whole mats, are all ways students represent subtraction situations.

Ideas for Instruction and Assessment

ACTING OUT OR USING MANIPULATIVES TO SHOW *TAKING FROM*

Students act out the problem and use concrete manipulatives to visualize *taking from* problems. The learning is active and ongoing to provide students with numerous opportunities to understand the operation of subtraction.

Act Out Taking From

Students learn about *taking from* by acting out problems in this whole-class lesson.

Before the lesson, prepare problems to be displayed on the whiteboard or on chart paper.

Pose the problem:

5 girls were dancing.
2 girls stopped dancing and went back to their seats.
How many girls are still dancing?

What is this problem about?

What are we trying to find out?

What could we do so we can see exactly what is happening in this problem? (Act it out.)

What should we show first?

Yes, let's have 5 girls come up and dance.

Now what should happen?

Okay, 2 girls stop dancing and go back to your seats.

So how many girls are dancing?

That's right! There are 3 girls still dancing.

Stand near the girls and count.

Let's count: 1, 2, 3. There are 3 dancing girls now.

We had 5 girls dancing. Then 2 girls stopped dancing, and then we had only 3 girls dancing.

5 take away 2 is 3.

Pose the problem:

Gail was holding 7 crayons.
Karl took 4 from her.
How many did she have left?

What are we trying to find out?

How should we start?

Have a student hold 7 crayons.

Now what happens?

Have Karl take 4 crayons.

What are we trying to find out?

How many crayons does Gail have?

Let's count: 1, 2, 3.

Can someone tell us what happened in the problem? So if you take 4 from 7, you get 3.

Tell students that they have been subtracting. They have been taking something from a group.

> To ensure that all students engage in the learning, prepare the problems ahead of time, making sure that all students have an opportunity to act in at least one problem.

> The focus of the lesson is on students acting out the problems, not reading and writing equations.

> Facilitate the engagement of the students who are not acting by asking them to use their fingers to show the problem as the actors are acting. (Begin by showing 5 fingers to represent the 5 dancers. Then take 2 away when the girls stop dancing.)

> Although this is not the wording we want students to use when they read subtraction equations, it is reasonable wording to use when we are initially building understanding of subtraction as taking from. We will transition to students saying 5 minus 2 is 3.

Have them talk to partners about what was *taken from* or *taken away* (e.g., crayons were *taken away*, dancers were *taken away*).

Pose and act out other *taken from* problems like the following:

> Use the names of students in your class in these problem situations.

Carlos had 3 cookies in his lunch.
He gave 1 to Kevin.
How many did he have left?

Betty had 5 books in her hands.
She gave 2 to Danny.
How many did she have left?

Have students act out each problem and talk about what was *taken away* or *taken from* the group (e.g., cookies were *taken from* Carlos and books were *taken from* Betty).

Taking From *with* Bears

Students use manipulatives to act out *taking from* subtraction problems.

> Modify the problems to match the manipulatives that you will be using for instruction.

Materials: bag of manipulatives for each student in a small group

It could be helpful to write the problems on the board or on chart paper for the students to refer to as they work.

> **Differentiation**
> Some students may benefit from −1 problems.

We have some story problems to solve about bears today.

There were 6 bears in the woods.
4 bears ran away.
How many bears were in the woods now?

Tell your partner what this problem is about.

Listen to this story problem to hear what we need to figure out.

Reread the story.

What do we need to find out?

How could we use the bear counters to solve the problem?

How did this problem start? How could you show that?

Show me the 6 bears.

Then what happened?

How could we show that?

Take 4 of your bears away because they ran away.

How will you know how many bears are left?

Count the bears that are left with me: 1, 2.

Let's see if what we did matches the problem.

Reread the problem.

There were 6 bears and you took away, or subtracted, 4 bears. Now you have 2 bears.

6 take away 4 is 2.

Pose the problem:

Max has 8 bears.

He takes away 4 bears to give to his friend.

How many bears does Max have now?

What is this story about?

How did the story start?

How could you show that?

Reread the story.

Then what happened?

What did you need to find out in the problem?

Turn and talk to your partner. Tell each other how you solved the problem.

So, how many bears does Max have now?

He started with 8 bears. Then he took away 4 bears.

Let's count what he has left: 1, 2, 3, 4.

Let's see if what we did matches the problem.

Reread the story.

There were 8 bears and you subtracted 4 bears. When you take away, you are subtracting. Now you have 4 bears.

8 take away 4 leaves 4.

Pose additional problems as needed.

Differentiation

As an extension, some students may want to create their own problems and act them out. They could talk about the problem and not have to write it.

As students continue to use manipulatives to act out taking from problems, give them time to solve problems independently. Then, ask individuals to explain how they solved the problem to the class. Their explanations will give you a clear indication of their level of understanding.

This student uses bears to show 7 − 2 as he takes 2 away from the group.

FORMATIVE ASSESSMENT

Pose a problem and give students manipulatives to use as they solve it.

I have 5 butterflies and give Sara 2.
How many butterflies do I have now?

Record observational notes of student work and responses.

Pose the problem to match the manipulatives you will be using.

Use the observational notes page.

Students should be able to show 5 manipulatives and then remove 2. They should recognize that the answer is the remaining objects and be able to count that there are 3 left. Watch to see if they know to first place the 5 objects to show the initial group.

Do they understand that the 2 should be removed? A common error is students adding 2 more counters to the group. Reread the problem if they add 2 more counters and ask them to tell you what is happening in the problem. Have them show you what it means to "give Sara 2."

Additional Ideas for Support and Practice

The following ideas extend students' understanding of using manipulatives to solve *taking from* problems and provide meaningful practice.

Monster Musical Chairs

Using a ten frame and magnets, or an interactive whiteboard, students represent the story *Monster Musical Chairs* by Stuart J. Murphy and then play musical chairs.

Before Reading:

Ask students:

Have you ever played musical chairs?
What happens when the music stops?

Tell the students that you are going to read a story about monsters playing musical chairs and learn what happens when the music stops.

During Reading:

Read *Monster Musical Chairs.*

Use a ten frame and magnets, or a ten frame and counters on an interactive whiteboard, to represent the story.

As you read each page, another monster is out of the game and a student can remove a magnet from the ten frame or slide a counter off the ten frame on the interactive whiteboard.

Discuss and have students retell what happened to the number of monsters after each page.

After Reading:

In small groups, students will have the opportunity to play musical chairs.

Set up 1 less chair than students in the small group.

Play music and have the students play musical chairs.

Whenever a student is "out," he or she must explain (using numbers) what has happened.

For example, there were 6 students, 1 student is out. Now there are 5 students.

Play continues until 1 student is left.

Partner Number Cube Game

Partners roll a number cube and take away cubes on a ten frame.

Materials for each player: 10 connecting cubes, a blank ten frame, and a number cube (1–6)

The game begins by placing all 10 cubes on the ten frame.

Player A rolls the number cube and takes away that many cubes.

Player A tells what is happening (e.g., if the number 4 was rolled, she would say, "There were 10 cubes, I took away 4 and now there are 6.")

Player B rolls the number cube and does the same thing on his frame, removing cubes and saying what happened.

Students compare the cubes left on their frames. The player with the most cubes on the frame wins that round.

Then they put the cubes back on and play again.

The game can be modified by placing different numbers of cubes on the frame at the start (e.g., 9 cubes for each player).

Notice the connections to comparing numbers and decomposing tens.

Ice-Cream Sundaes

Students solve ice-cream problems using pom-poms.

Materials: a plastic bowl and plastic spoon for each student; large craft pom-poms (cotton balls will also work), empty gallon ice-cream bucket (optional) for each group

In a small group, pass out the bowls and spoons to each student.

The pom-poms should be in the plastic ice-cream bucket ready to "serve."

Begin by telling a subtraction story (e.g., Jane ordered a 7-scoop ice-cream sundae. She ate 4 scoops. How many scoops does she have now?)

Students retell the story and then solve by using the pom-poms as ice-cream scoops (each pom-pom is a scoop of ice cream).

They pretend to be Jane and "eat" the ice cream with their spoons.

When solved, ask students to explain the process and the solution.

Repeat with a new ice-cream story.

DRAWING TO SHOW *TAKING FROM*

Students draw simple pictures to illustrate the story problems.

The pictures do not have to be detailed because the focus of a math picture is to show what is happening in the problem. The pictures can be as simple as circles or squares to represent the problems.

Drawing Pictures to Show Taking From

Students will draw pictures to solve *taking from* subtraction problems.

For this whole-class lesson, write problems on chart paper and allow space for drawing pictures to solve the problems.

Pose the problem:

Manny had 9 cookies.
He ate 3 cookies.
How many cookies does he have now?

Turn and tell your partner what is happening in this story.
What do we need to find out in this problem?

Reread the problem.

Turn and tell your partner how you would solve it.
Show me how you would solve it with a picture.
What would you draw?
Why should I start with 9 cookies?
What happened next in the problem?
In my picture, how can I show that he ate 3 cookies?
I could cross them out. I don't want to erase them because I don't want to forget about them.
Cross out the cookies Manny ate. Think about how many he ate.
So how many cookies does Manny have now? How do you know?
Let's count: 1, 2, 3, 4, 5, 6.
Does what we did match the problem?

We are exposing students to the language and notation of subtraction. They do not immediately have to use this language or write equations, but they gain comfort and familiarity with it as they see and hear it.

Reread the problem.

Manny had 9 cookies. He ate 3, and now he has 6 cookies.
9 take away 3 is 6.

Record $9 - 3 = 6$ on the board.

We say 9 minus 3 is 6.

Pose the problem:

Emma has 10 stickers.

She gave 3 stickers to her sister.

How many stickers does Emma have now?

Turn and tell your partner what is happening in this problem

What are we trying to find out?

Reread the problem.

Draw a picture to show Emma's stickers.

I'm going to draw circles to show the stickers, but you could draw squares or rectangles or whatever you'd like.

How many circles did you draw to show the stickers? Explain.

Draw 10 circles on the board.

Emma gave 3 stickers to her sister.

How will you show that in the picture?

Cross out 3 circles.

How many stickers does Emma have now? How will you know?

Go ahead and count them. How many does she have left?

Does what we did match the problem?

Reread the problem.

I have 7 circles, or stickers, left in my picture.

She had 10 stickers and gave away 3. Now she has 7 stickers.

Write $10 - 3 = 7$ on the board.

10 minus 3 equals 7.

Turn and tell your partner why we subtracted to solve this problem.

How did we show subtraction in our pictures?

Drawing pictures to solve problems should be ongoing in whole or small groups. As students become proficient with drawing pictures, they can write the numbers shown as labels. It is an important foundation for later writing equations to represent the problems.

FORMATIVE ASSESSMENT

Give students a piece of paper for them to draw a picture to solve a problem.

10 frogs were sitting on the log.

2 frogs hopped away.

How many frogs are on the log now?

Record observational notes of student work.

Use the observational notes page.

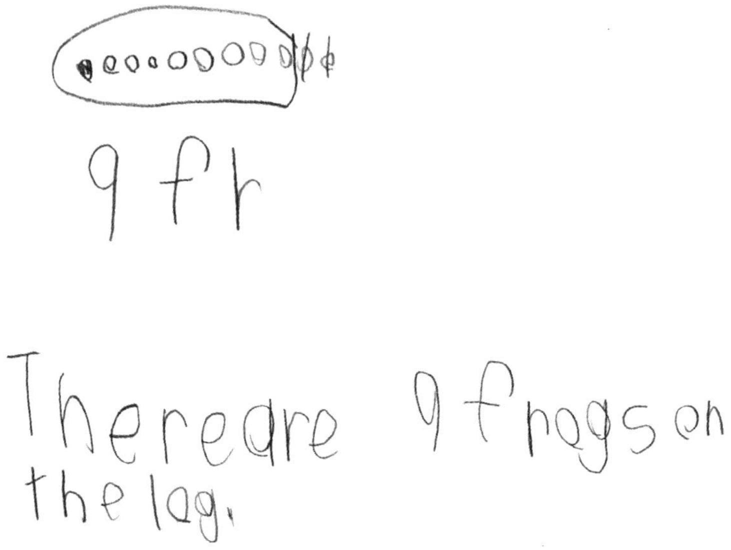

This student does not accurately draw a picture to represent the 10 frogs. He draws 11 dots. He takes away 2 frogs but is left with 9 frogs. He labels his answer and writes a sentence to match the problem. Ask the student what the dots represent (frogs) and how many there were (10). Have him count the dots for you to see if he can catch his own mistake.

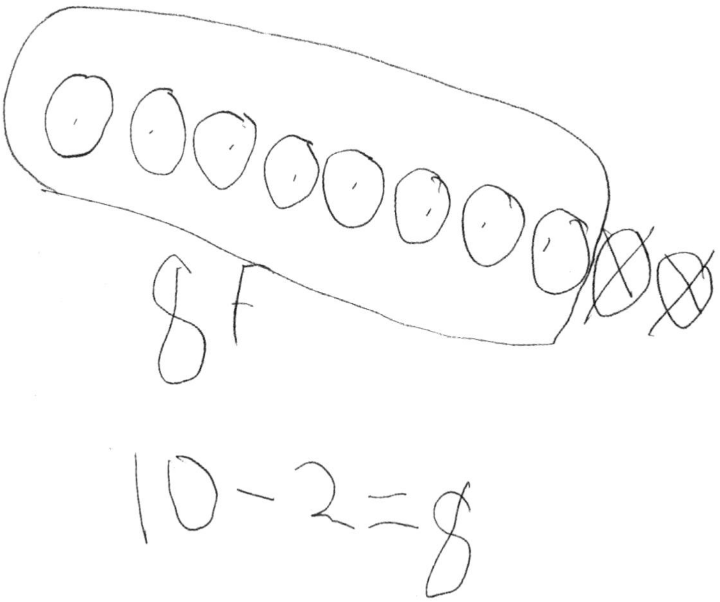

This student accurately draws a picture to solve the problem. He draws 10 circles and crosses out 2 to show the frogs that hopped away. He counts the number of frogs that remained by putting a dot in each circle. Then he labels the answer with *8 F* for 8 frogs. He also writes the number sentence to match his picture.

Additional Ideas for Support and Practice

The following ideas extend students' understanding of drawing pictures to solve *taking from* problems and provide meaningful practice.

Pictures on Ten Frames

Students solve problems using ten frames in sheet protectors to draw pictures.

Materials: ten frame in a sheet protector and a dry erase marker for each student

Pose a problem to solve.

The clown had 3 balloons.
1 of them popped.
How many balloons are there now?

Students draw simple pictures on the ten frame to solve.

Then they write the answer below the picture.

Continue with other problems.

There were 3 kites. 1 flew away. How many are there now?

There were 4 dogs. 2 ran away. How many are there now?

There were 3 lollipops. I ate 1. How many are there now?

Picture Posters

Students create posters with pictures to solve problems.

Fold pieces of construction paper into three parts.

Write some problems for students to solve or use some problems from the online resources.

Students select one problem to solve and cut it out.

In the first section, students glue the problem.

In the middle section, students draw a picture to solve the problem.

In the third section, students write a number sentence to match the picture.

Students share their posters and explain how they solved the problem.

This activity can be extended by folding the paper into four parts.

In the fourth section, students can write a sentence to explain how they solved the problem.

ACTING OUT OR USING MANIPULATIVES TO SHOW SEPARATING

Although taking apart is a subtraction structure (There are 6 children. 2 are girls. How many are boys?), in these initial explorations of taking apart, students are asked to find both parts (There are 6 children. How many are girls? How many are boys?). When one part is known, we see the subtraction (6 – 2), but when neither part is known, it feels more like an addition problem because we are looking for all of the ways 6 can be separated (6 = 2 + 4 or 6 = 3 + 3 or 6 = 5 + 1 . . .). At this simplified kindergarten level, students represent this with addition equations.

Students explore *taking apart*, or *separating*. This problem-solving structure connects to decomposing numbers in Module 6.

Acting Out Separating

Students act out problems about boys and girls in line to show *taking apart*.

Tape a blue paper at one end of the board and a red paper at the other end.

Line up 6 students in front of the class between the two colored papers.

How many students are there? (6.)

Let's separate these students into two teams, a blue team and a red team.

Have 2 students stand below the blue paper and 4 stand below the red paper, leaving space in between the teams.

How many students are on the blue team?

How many students are on the red team?

Are there still 6 students?

Record on the board $6 = 2 + 4$.

Have the students move back together.

Do you think there is another way we could separate these students into two teams?

Allow students to suggest how many students might be on the blue team and model the new teams (e.g., 3 on blue and 3 on red).

How many students are there altogether?
How many students are on the blue team?
How many students are on the red team?
Are there still 6 students?

Record on the board *6 = 3 + 3*.

Have the students move back together.

Do you think there is another way we could separate these students into two teams?

Allow students to suggest another possibility (e.g., 5 and 1) and then model it with them.

Record *6 = 5 + 1* on the board.

Have students look at the equations on the board.

Have we done something like this before? When? (It's like when we broke cubes to see how many ways we could make 6.)
Was there more than one way to separate the cubes?
Does it make sense that there is more than one way to separate 6 students?

Act out splitting 7 students into the blue and red teams.

Continue to write the equations and talk about the different ways the 7 students can be put into two teams.

Thinking Through a Lesson

Separating with Connecting Cubes

Students explore the concept of *separating* by building and separating cube caterpillars into parts.

Materials: a bag of 8 red and 8 yellow connecting cubes for each student

We are going to make red and yellow caterpillars.
Build your caterpillar using a total of 5 cubes.
You can use any 5 cubes you'd like.

SMP1, SMP4

Have students pick the 5 cubes they want to use, put the red together, then the yellow together, and then connect them to make their caterpillar.

You may need to demonstrate placing the red cubes together and yellow cubes together (e.g., R, R, R, Y, Y rather than R, Y, R, Y, R). Having the colors together makes the parts more visible to students.

As students are building their caterpillars, make sure they are using 5 cubes and are using both red and yellow cubes.

Give students time to build their caterpillars.

What is the same about all of our caterpillars? (They are 5 cubes long.)

Do you think we all used the same number of red and yellow cubes? (No.)

How could I find that out? (Count the red and yellow.)

Separate your caterpillar into parts: red and yellow.

Students break their caterpillars into red and yellow parts.

Count how many red and how many yellow cubes you used.

Have students show their two caterpillar parts to partners and tell them how many red and yellow cubes are in their caterpillars.

Do you have the same number of red cubes and yellow cubes as your partner?

Do you have the same total number of cubes? (Yes, 5.)

Have some students share with you as you record the possibilities on the board.

Derek used 1 red and 4 yellow.

How many cubes were in his whole caterpillar? (5.)

Have students who used those cubes hold up their cubes.

SMP2

The notation for this is addition at this level. This connects to what students have done in decomposing numbers. Although we are showing it as addition at this level, it is developing the foundational thinking that will become subtraction in first grade when students are given one part and must find the missing part.

Record $5 = 1 + 4$ on the board.

Colin used 4 red and 1 yellow.

How many cubes were in the whole caterpillar? (5.)

Have students who used those cubes hold up their cubes.

Help students see that although these numbers are the same, the caterpillars with 4 red look different from the caterpillars with 1 red.

Record $5 = 4 + 1$.

Continue to ask students to share as you record how they separated the caterpillars

Record $5 = 2 + 3$.

But some have 3 red and 2 yellow cubes.

Record $5 = 3 + 2$.

We showed lots of different ways to separate our caterpillars.

Put your caterpillar back together. How many cubes is it? (5.)

We separated 5 a lot of different ways, but they are all 5.

By putting the caterpillars back together, you are showing the connection between putting together and taking apart (separating).

Have students build another red and yellow caterpillar.

This time your whole caterpillar will have 7 cubes.

Part of the caterpillar will be red, and the other part will be yellow.

Give students an opportunity to build their 7-cube caterpillars, then have them split their caterpillar into parts and count the number of cubes in each part.

When they are finished, give each student an opportunity to explain *how* they separated their caterpillar, asking them to use the words *part* and *whole* as they talk about their caterpillar.

Record the possible ways the caterpillar could be separated.

If time permits, repeat with a 6-cube caterpillar.

Have students draw two caterpillars that are 8 cubes long in their journals.

Have them color them to show two different ways they could separate them.

Have them record the parts by completing 8 = __ + __ for each caterpillar.

Asking students to use the words part and whole focuses them on the key ideas of the lesson and helps them explain their thinking with greater precision.

SMP4

Differentiation

Students can be challenged to show as many ways to break apart the caterpillars as they can find.

Problem Solving with Number Bonds

Students revisit number bonds to explore *take apart* problems.

Pose the problem:

There were 6 fish in an aquarium.
Some had spots and some had stripes.
How many of each type could there be?

Turn and tell your partner what this story problem is about.
How many fish are there?
What are you trying to find out?

Draw the following number bond on the board.

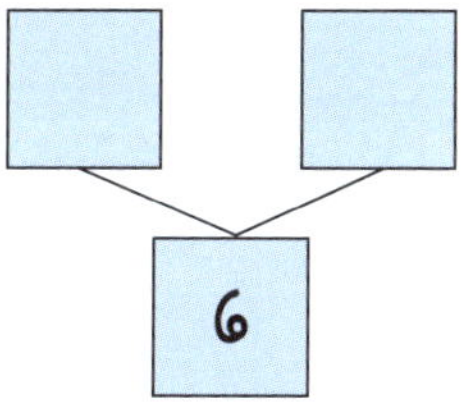

What is this?

Could we use it to show what is happening in this problem?

Turn to your partner. What do you think this 6 shows?

How is a number bond like a part-part-whole mat?

Is the 6 in our problem a part or the whole?

Reread the problem, then allow students to talk with partners.

What is the 6 in this problem?

So what is in the other squares?

What could we label them? What are the parts in this story?

Label the squares *spots* and *stripes*.

Talk with your partner and come up with a possible answer. How many fish had spots and how many had stripes? You can use the counters at your table if you'd like to see it that way.

Have students share their answers and fill in the number bond, then create additional number bonds to show the different answers (3 and 3, 4 and 2, 5 and 1 . . .).

Did we solve the problem?

Reread the problem.

Was there just one answer?

What do you notice about the number bonds? (They all have 6 as the whole.)

Pose the following problem:

> A farmer planted yellow squash and green squash.
>
> He planted 10 squash.
>
> How many could be yellow and how many could be green?

Turn and tell your partner what this problem is about.

How many total squash did he plant?

What do we need to find out?

You may use counters or use a number bond or use both to find an answer to this problem. Work with your partner to solve this problem.

Have students share their solutions, having those that used counters share first.

Then, invite students who made number bonds to show them. Have them tell how they knew where to put the 10.

Show a few possible solutions.

Reread the problem to see if the solutions match the problem.

FORMATIVE ASSESSMENT

Pose the problem:

> Jenny had 7 apples.
> Some were red and some were yellow.
> How many of each?

Ask students to show you two possible solutions. Give them 2-color counters that they may use to show the problem.

Students should be able to show at least two ways of *separating* 7 into two parts (e.g., 5 and 2 or 3 and 4). Have them explain how they figured out the answers (maybe counting out 7 counters and separating them into two groups). Some students may solve it without counters (e.g., by simply knowing that 6 and 1 make 7).

Record observational notes of student work and responses.

If they are unable to solve the problem, ask students to show you all of the apples. Have them tell you what they know about the colors of the apples (some are red and some are yellow) and what they are trying to find out (how many of each).

Use the observational notes page.

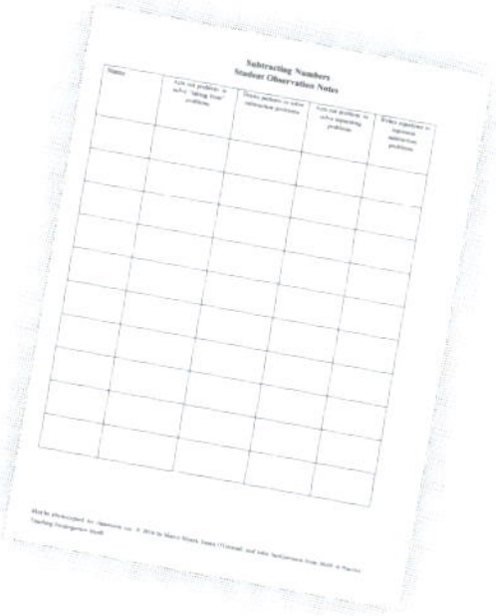

Additional Ideas for Support and Practice

The following ideas extend students' understanding of using manipulatives to separate numbers and provide meaningful practice.

Student-Created Problems

Students use part-part-whole mats while creating problem-solving stories.

Materials: a part-part-whole mat and counters for each student (students can work independently or with a partner for this activity)

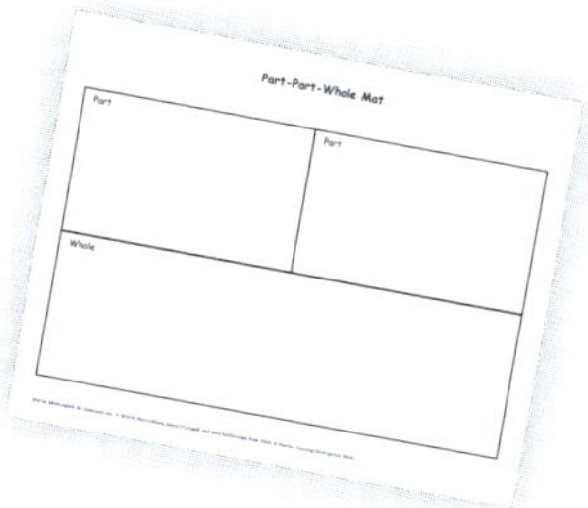

First, ask for an amount and an idea for an object (for example, 7 chairs).

Then, ask the students to give you two descriptors of the objects (for example: big/small or round/square).

Next, students help create stories about the named objects.

> *Let's start our story with 7 chairs. What can you say about the parts? Good idea! There were 7 chairs. Some were big, and some were small.*

Finally, students show one possible answer on their part-part-whole mat.

The teacher then represents the problem using a number bond when students find an answer.

REPRESENTING SUBTRACTION WITH EQUATIONS

Students begin to connect equations to the concrete and pictorial ways they have been solving problems.

Writing Equations for Subtraction

> Students need to have an understanding of the concept of subtraction before writing equations. They need to be able to model subtraction with manipulatives and pictures before moving on to the abstract representations of numbers and symbols.

Students solve problems using manipulatives or drawing a picture and then learn to write subtraction equations to match the problems. This lesson should take place after students demonstrate an understanding of *taking away* and *separating* numbers.

Have counters, paper, and pencils available at tables for students to use as they solve the problems.

Pose the problem:

There are 6 turtles on a log.
2 turtles swim away.
How many turtles are on the log now?

Turn and tell your partner what is happening in this problem.
Show me how you can solve it.

Observe as students solve it.

How many turtles are on the log now?
How did you solve it?
There were 6 turtles on the log and 2 swam away. Now there are 4 turtles.

Record $6 - 2 = 4$ on the board.

Turn and talk with your partner. What is the 6?
What is the 2?
Why did I write this (–)? What does it mean?
We wrote a number sentence to show what happened in that story.

Reread the problem.

Does our number sentence, our equation, match the story? How?
I wrote 6 minus 2 equals 4.

Point to the minus sign.

6 minus 2.
What does minus mean?

Point to the equal sign.

What does that mean? (Equal. *Or*, The same as.)

Have students write the number sentence.

Pose the following problem:

8 birds were in a tree.
3 birds flew away.
How many birds are in the tree now?

Have students turn and tell partners about the problem and what they are trying to find out.

Should we use addition or subtraction to find that out? Explain.

Have students solve the problem with partners and then share their answers and methods.

What number sentence can we write to show what happened in the problem?

Ask students to share a possible equation as you record it on the board, asking them to tell what each number and symbol represents.

Does our number sentence match our story?

Reread the problem.

Now write the number sentence near your picture or on your paper.

> *This type of lesson should be repeated as students continue solving problems throughout the year. It is important in kindergarten that students use manipulatives and draw pictures to solve problems. They may also write number sentences or equations to go with the representations.*

FORMATIVE ASSESSMENT

Option 1

Have students write a subtraction equation for each drawing.

This student shows her understanding of the pictures as subtraction and is able to accurately build an equation for each model.

Common Error

A common error for the first picture is writing 4 - 1 rather than 5 - 1. Students often forget that what was crossed off is part of the whole.

Look for students who build equations first showing the number that represents the total number of suns, then the −1 to show what is crossed off, and finally the difference.

In this work sample, the numbers 3 and 2 are reversed, which is common for students in kindergarten. Reminding the student of the correct number formation is important, but in this case, it is evident which numbers the student intended to write. Placing a number strip near her table for reference may be helpful, as would taping a number card at her table to show the formation of any numbers with which she has particular difficulty.

Option 2

Give students a problem to solve by drawing a picture and writing an equation to match.

> The baker made 6 cupcakes.
> Then she ate 2 cupcakes.
> How many cupcakes are there now?

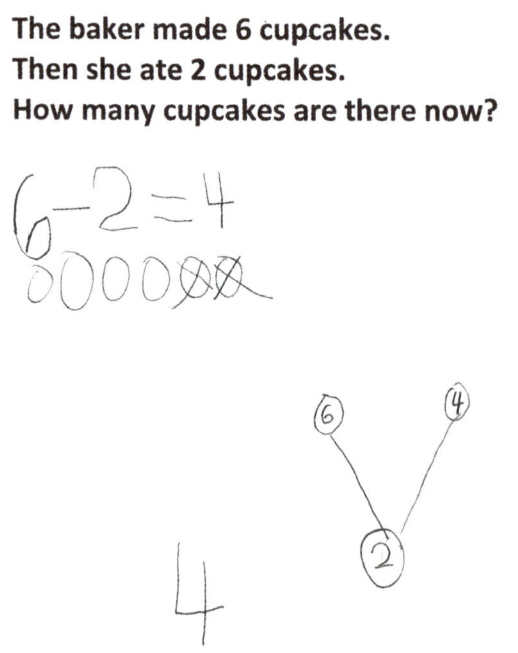

This student draws 6 circles and crosses out 2, showing his understanding of the problem. He correctly builds the equation to show the problem as 6 − 2 = 4. The student also tries to represent the problem with a number bond but incorrectly shows the number bond as a path from 6 to 2 to 4 rather than 6 as the whole and 2 and 4 as the parts. Because this student appears to understand the problem, it is likely he simply needs to revisit the way in which a number bond is drawn to show the parts and whole.

Using Pinch Cards to Show Addition or Subtraction

Students will listen to story problems and use pinch cards to show if the problem is addition or subtraction.

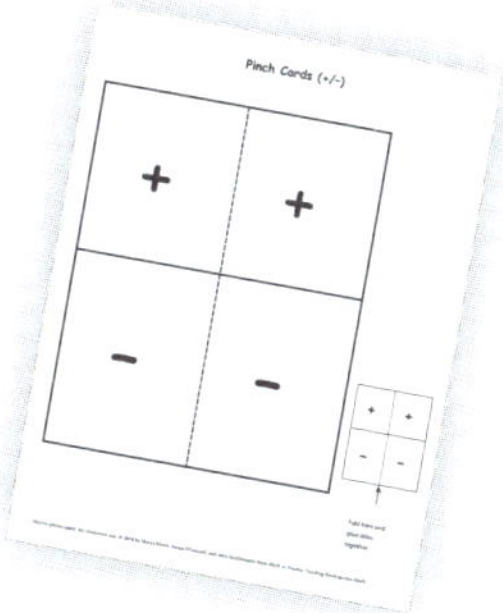

Materials: an addition and subtraction pinch card for each student

Have problems prepared on chart paper or an interactive whiteboard for this whole-class lesson.

Here's a pinch card for each of you.

What do you notice about it?

We are going to use these as you listen to story problems.

Pinch the plus if you think the problem is solved with addition. Pinch the minus sign if you think it is solved with subtraction.

Here's the first problem:

Hailey has 8 pieces of candy.

She eats 5 of them.

How many pieces of candy does Hailey have now?

Think about that problem.

Do you think you would add or subtract to solve the problem? Show me with your card.

Hold up your card when you know.

I see many of you showing the minus sign. Why do you think this is subtracting?

Taking some away is subtraction.

A student uses a pinch card to show that the problem is addition. Pinch cards are a great way for all students to respond and for teachers to quickly see their responses.

Here's another problem:

Liam has 4 apples and 2 oranges.

How many pieces of fruit does Liam have?

Would you add or subtract to find out how many pieces of fruit he has? Pinch the plus or minus sign.

Turn and show your neighbor. Tell your neighbor what type of problem it is and how you know.

I see some people think it is addition. Why?

Continue with other addition and subtraction problems.

Additional Ideas for Support and Practice

The following ideas extend students' understanding of writing equations for subtraction and provide meaningful practice.

Writing Equations to Match Pictures

Students write equations to match picture cards.

Materials: large index cards with up to 10 stickers, some stickers crossed out

Students select a card, write the equation for the problem on a sticky note, and put the note on the card. For example:

A card shows 7 heart stickers with 3 crossed out.

The student writes $7 - 3 = 4$.

The student's work can be assessed by quickly reviewing the cards.

Talk About It/Write About It

What does it mean to subtract?

Draw a picture tshow 6 − 1.

There are 5 spiders. What could happen to show a subtraction problem?
Draw a picture to show the subtraction problem.
Write the number sentence to match the picture.

There are 9 students on the bus. What could happen to show a subtraction problem?
Draw a picture to show a subtraction problem.
Write the number sentence to match the picture.

Partner Problems

Students work with a partner to make up problems and write equations to solve them.

Write numbers 0–10 on large craft sticks, and place them in a cup.

On another set of large craft sticks, write simple words that students can use in problems (*dogs*, *cats*, *dolls*, *cars*, *books*). Place these sticks in a different cup.

Player A selects two number sticks and one word stick to make up a problem (e.g., Player A chooses 5, 2, and dolls).

She says: "Harriet has 5 dolls. She gives 2 dolls away. How many dolls does she have now?"

Player B draws a picture to solve the problem.

Player A writes a number sentence to solve the problem.

They return the sticks to the cups.

Players switch roles and do it again.

Vocabulary

equal (same as)
minus
separate
subtract
symbols (−, =)
take apart
take away
take from

Online Resources

General Resources

Additional Problems

Part-Part-Whole Mat

Pinch Cards (+/−)

Subtracting Numbers Student Observation Notes

Ten Frame Mats

Resources for Specific Activities

Cupcakes Problem Sheet

Frogs Problem Sheet

Subtraction Recording Sheet

These resources are available at http://hein.pub/MathinPractice, keycode MIPGK.

MODULE 9

Understanding Math Facts

About the Math

K.CC.4c; K.CC.2; K.CC.6

Kindergarten students learn about the operations of addition and subtraction. They learn to represent the operations both through pictures and through numbers and symbols ($2 + 1$ or $2 + 1 = 3$). At this level, the focus is on understanding these operations, not memorizing sums and differences.

There are several basic understandings about numbers developed during the kindergarten year that support students' initial work with addition and subtraction facts. These understandings include:

- recognizing that in the counting sequence each number is 1 more than the number said before it
- counting on from any given number in the counting sequence
- the concepts of same, more, and less.

Connecting the math facts of +/−1 and +/−0 to these number concepts allows kindergarten students to make sense of these fact sets.

The key ideas focused on in this module include:

- exploring +/−1 to make connections to 1 more and 1 less
- exploring +/−0 to discover that adding or subtracting 0 does not change a quantity
- writing equations to show +/−1 and +/−0 facts.

EXPLORING +1

Students explore the concept of 1 more during their study of numbers. They build numbers and then show 1 more with ten frames and towers of cubes (see Figure 9.1). They pick a domino to show a number and then find a domino that shows the number that is 1 more. They find a number on the hundred chart and then circle the number that is 1 more. Connecting 1 more to the concept of adding 1 clarifies the meaning of +1 (increasing the quantity by 1).

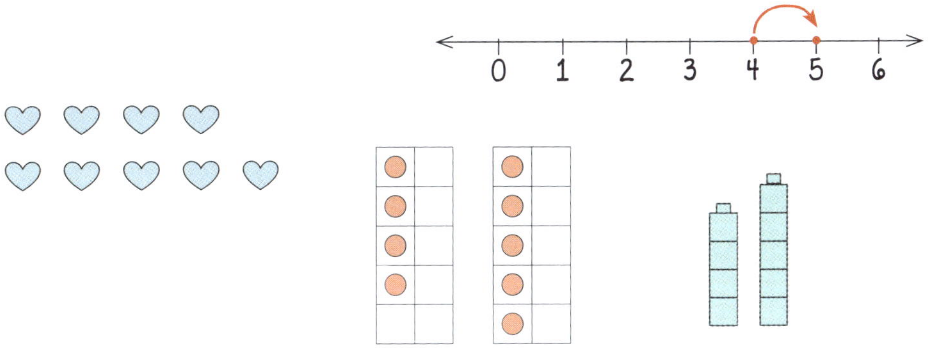

Figure 9.1 Students have many prior experiences visualizing 1 more and now connect that idea to the concept of adding 1.

Kindergarten students practice counting on from any number. When they combine that skill with the understanding that adding 1 to a number results in the next number in the counting sequence, there is really no need to memorize +1 facts. Being able to solve +1 facts from 1 + 1 to 10 + 1 is a matter of understanding, not memory. At this level we focus on __ + 1 to allow students to use their *1 more* or *plus 1* understanding. In first grade, they explore 1 + __ and discover the commutative property.

EXPLORING -1

Students also explore the idea of *1 less* as they work with numbers 1–10. Through stories, hands-on activities, and drawings, students connect *minus 1* to their understanding of *1 less* (see Figure 9.2). Modeling the expressions (3 − 1) and equations (3 − 1 = 2) helps students see the symbolic way we show subtraction. Knowing the answer to 5 − 1 is not about memory, but about understanding the number that is 1 less, or that comes just before that number in the counting sequence. At this level, we focus on __ − 1 only, rather than subtraction problems that have a difference of 1 as in 7 − 6 = 1. Those facts are explored in first grade.

The grade-level fluency progression in this series may vary slightly from the grade-level expectations in other standards, but the ultimate goal of addition/subtraction fluency by the end of second grade remains the same.

Learning Goals

I can add 1 to numbers to 10.

I can subtract 1 from numbers to 10.

I can add 0 to numbers to 10.

I can subtract 0 from numbers to 10.

Lessons in This Module

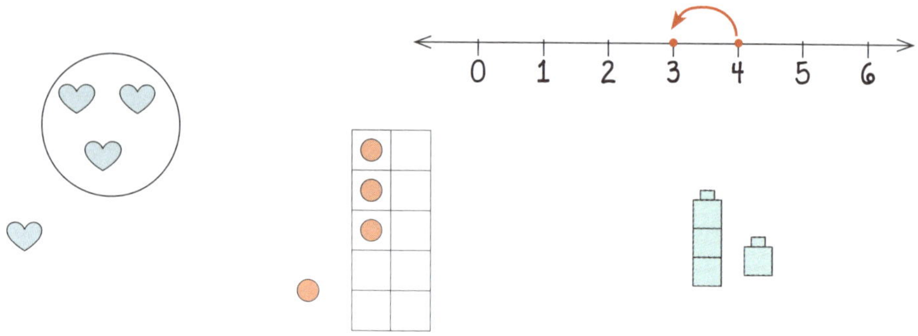

Figure 9.2 Students explore −1 in hands-on and visual ways, making connections to their understanding of 1 less.

EXPLORING +0

Through discussions of the value of 0 and investigations in which we add nothing to a set of cubes or counters, students begin to see that adding 0 does not change a quantity.

- As students place 5 counters on a ten frame and then add no more counters, they see that they still have 5 counters.
- As students use part-part-whole mats to show 4 counters and 0 counters, they see that the total is 4 counters.
- As students build a tower of 6 cubes and then add no more cubes to their tower, they see that the tower is still 6 cubes tall.

Once students understand this concept, there is no need to memorize these facts. They are built on the understanding of 0.

EXPLORING -0

Through discussions of the value of 0 and investigations in which we take nothing from a set of cubes or counters, students begin to see that subtracting 0 does not take anything from a number.

- As students gather a set of 5 blocks and then take no blocks away, they see that they still have 5 blocks.
- As 6 students stand to get in line and no students go back to their seats, they count and see that there are still 6 students in the line.

Memorization is not necessary for students to know −0 facts.

Beginning the fluency process in kindergarten helps us to ultimately meet the goal of fluency with all addition and subtraction math facts by the end of second grade. Fluency at this level, however, is not dependent on memorization. If the facts that kindergarten students are asked to commit to memory are based on their understanding of the concepts of 1 more, 1 less, and the same (our +/−1 and +/−0 facts), we are actually expanding their number sense as they begin their work on fact fluency. Memorizing arbitrary math facts at this level is counterproductive. The math facts students commit to memory should make sense with the math concepts they are exploring.

Fluency activities at this level are tasks that give students repeated opportunities to apply their number understandings to math facts. These activities are designed to reinforce the number concepts they have been exploring all year. Students roll dot cubes or pick ten frame cards and identify the number represented by the dots, then add or subtract from it. Students tell partners the quantities they see and practice writing equations to show addition and subtraction. Students visualize addition and subtraction as they build on ten frames, connect cubes for towers, draw pictures to show the facts, and write the equations.

Through stories, investigations, discussions, and interactive tasks related to *1 more*, *1 less*, and *same as* (+/−1 and +/−0), our students gain understanding of and fluency with some basic fact sets prior to first grade. First-grade teachers are then able to continue the progression of math facts to refine students' understanding of +/−1 and +/−0, as well as build understanding and fluency with additional fact sets.

Understanding +0
- Adding 0
- +0 on Ten Frames

Developing Fluency with +0
- Fact Bracelets
- +0 Memory
- Counting +0 or +1
- Dot Cards +0 Partner Game

Understanding −0
- *1, 2, 3 to the Zoo*
- *Writing −0*

Developing Fluency with −0
- Hopscotch
- Ten Frame Subtraction
- Talk About It/Write About It
- Partner Towers
- Spin and Draw −0

Exploring the Progression

PREVIOUS		NOW		NEXT
Preschool	»	**Grade K**	»	**Grade 1**
No prior experience with addition and subtraction facts		Exploring adding and subtracting 0, 1; writing equations to show +/−1 and +/−0 facts		Extending understanding of +/−1 and +/−0 facts; building fluency with additional fact sets

Ideas for Instruction and Assessment

UNDERSTANDING +1

Students connect the idea of adding 1 to the concept of 1 more as they practice counting on from any number.

Using Ten Frames to Show +1

Students explore adding 1 with ten frames. They connect adding 1 to finding the next number in the counting sequence and the concept of 1 more.

Each student needs 10 counters and a ten frame mat.

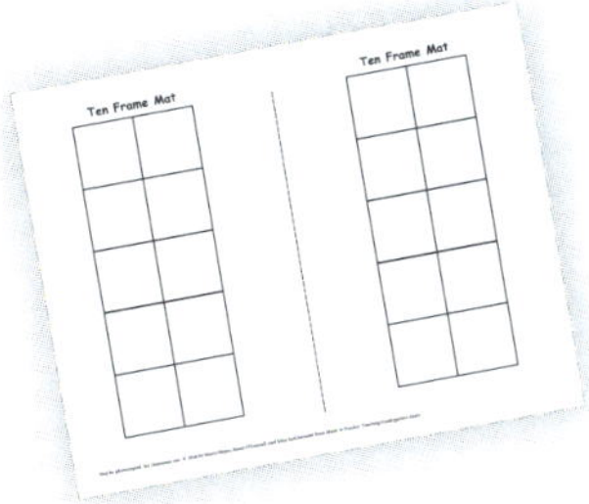

Pose the problem:

> There were 6 dogs at the dog park.
> 1 more dog ran into the park.
> How many dogs were at the park then?

How many dogs were at the park when the story started?

Use your ten frame to show me the dogs at the park.

What happens next in the story?

Show me that on your ten frame.

How many dogs are at the park then?

How do you know?

Turn and tell your partner how we could show that with a number sentence, an equation.

Record *6 + 1 = 7* on the board.

What is the 6?

What is the 1?

What is the 7?

Why did we add?

Notice the connection with the problem-solving lessons in Module 7. Beginning with problems allows students to make sense of what 6 + 1 means.

Have students clear their mats.

Pose the following:

There were 4 squirrels in the tree.

1 more squirrel scurried up the tree.

How many squirrels were in the tree then?

How many squirrels were in the tree when the story started?

Use your ten frame to show me the squirrels.

What happens next in the story?

Show me that on your ten frame.

How many squirrels were in the tree then?

How do you know?

Turn and tell your partner how we could show that with a number sentence, an equation.

Record *4 + 1 = 5* on the board.

Have students tell what each number in the equation represents.

Why did we add?

Have students clear their mats.

Have students model and discuss several more +1 problems following the same line of questioning, recording the equations after each problem is solved.

6 + 1 = 7

4 + 1 = 5

2 + 1 = 3

7 + 1 = 8

Differentiation

Some students may need to spend more time adding 1 to the numbers 1–5 and others may quickly be able to move to adding 1 to the numbers 6–10.

Then, ask students to look at the equations listed on the board.

What do you notice?

What is happening when we add 1? (It's 1 more. Or, It's the next number we say when we count.)

Is that always happening?

Does that make sense? Explain.

So if you want to add 1, what could you do?

Pose 8 + 1.

How many counters do you think you will have when you add 8 + 1?

Listen for students' predictions. Have them explain them.

Then, have students do it to check their predictions.

Was it the next number? Were you right?

Turn and tell your partner: When you add 1, how do you know what the sum will be?

Are students using their observations and their conjectures to predict the next number?

Repeat with several more expressions if students need more time to think about it or more examples to test their thinking.

Plus 1 Towers SHOW IT!

In earlier lessons, students built towers to learn that each successive number in the counting sequence is 1 more than the previous number. This lesson revisits building towers to connect the concept of 1 more to the concept of +1 and focuses on the written expressions (6 + 1) and equations (6 + 1 = 7).

Each student in this small-group lesson needs 10 connecting cubes of one color and 1 cube of a different color to show +1. Students record their equations on dry erase boards or paper.

Give students 10 cubes of the same color.

Show me 6.

Give each student 1 more cube of a different color.

How many cubes do you think you will have if you put 1 more on your tower? Explain.

How can you use your cubes to show 6 + 1?

Observe as students create their towers.

Explain your 6 + 1.

How many cubes do you have altogether?

Was your prediction correct? Is that how many you thought you would have?

Think about how your tower shows +1. How could we write that in a number sentence?

Write the number sentence on your board to match your tower.

Students should connect 6 cubes and put the +1 cube at the top or bottom. In that way it is easier for students to see the initial quantity and the one that has been added. When the 1 cube is placed in the middle, it can be difficult for students to see the cubes as 6 + 1.

Have students show their equations. Discuss the numbers and symbols.

Record 6 + *1* = *7* on the board.

What is the 6?

What is the 1?

What does the equal sign mean?

What is the 7?

Pose a few more +1 expressions for students to predict, model with their cubes, and write and explain the equations.

Now, you get to build your own +1 tower and write a number sentence to match it. You can start with any number you want.

Observe that the number sentences match students' models.

Each time, have students predict how many they will have before adding 1 more cube, revisiting the idea of +1 being the next counting number.

Have students share their models and equations.

How does your number sentence match your tower? (It shows 1 more. *Or,* It shows +1.)

What do you notice about +1?

FORMATIVE ASSESSMENT

Interview:

What happens when you add 1 to a number?

What does it mean to add 1?

Use the observational notes page.

Listen for students telling you that adding 1 makes it the next counting number or that it is like having 1 more.

Record notes of student's responses.

If they are not able to verbalize this, try it again with a specific number and manipulatives.

Place 5 counters on the table.

Add 1.

What happened when you added 1?

What did you get when you added 1? (6 because it is 1 more than 5. *Or,* 6 because it is the next number.)

We are not looking for the number 6, but the insight that it is 1 more. 6 is correct, but doesn't convince us they have the big idea.

Place 3 counters on the table.

Add 1.

What happened when you added 1?

What did you get when you added 1? (4 because it is 1 more than 3. *Or,* 4 because it is the next number.)

If students are unable to do this, they are in need of more hands-on experiences in small groups similar to those in the preceding Plus 1 Towers activity. By observing and discussing what happens when they add 1 more to towers, add 1 more counter to ten frames, or add 1 more link to chains, they are beginning to get the big idea of what +1 means.

DEVELOPING FLUENCY WITH +1

These fluency activities give students repeated practice applying their number understandings as they work with partners or independently to practice facts.

Partner Towers

Students use ten frames to show +1 facts and write equations with partners.

Player A puts counters on a ten frame to show any number.

Then, he adds a +1 counter of another color.

Player B writes the equation to match.

Then, players switch roles.

These students practice +1 facts. He uses a foam ten frame and two-sided counters to show 4 + 1 = 5. She writes the number sentence to match. They will switch jobs to continue practicing.

Fish Eyes

Fish Eyes: A Book You Can Count On tells the story of a fish that wants to join other schools of fish and sets a context for +1 discussions.

Before Reading:

Do you think fish like to swim alone or in groups?

We are going to read a story about a fish that likes to swim with other fish.

During Reading:

Read *Fish Eyes: A Book You Can Count On* by Lois Ehlert.

After Reading:

Reread the story and have students use counters or goldfish crackers to show each set of fish, and then show the +1 fish that joins the group.

Have them record the +1 fact for each set of fish.

Have them observe the facts and tell you what they notice.

Additional practice for a center:

Students roll a number cube and show that number of fish using goldfish crackers or counters.

Then they add 1 more fish and write the +1 facts to show that number of fish.

Number Cards and +1 Equations

The teacher holds up a number card and students practice writing +1 equations.

Students have dry erase boards and markers.

The teacher holds up a number card (1–9) and says, "Plus 1."

The students write the correct answer and hold up their board.

When students are finished writing, a student is called on to say the equation (__ + 1 = __).

Allow all students time to write the equation so it does not become a race to finish first. It is more important for all students to be accurate than for a few to be fast.

BINGO

Students create bingo cards to practice +1 facts.

The teacher prepares the game cards by folding a paper into 3 × 3, or 9, spaces.

Students write numbers 2–10 in the empty spaces.

The numbers do not have to be in order.

The teacher pulls a number card (1–9) and says the number and "plus 1" (e.g., the teacher pulls a 5 card and says, "5 + 1").

Students cover the answer with a cube.

Play until a student gets three in a row for bingo.

Clear the cubes and play again.

I Say—You Say 1 More

Students practice +1 facts verbally whenever there are a few minutes available.

The teacher selects a number card 1–9 and says the number and "plus 1."

Students respond with the number that is 1 more.

This activity can be practiced while lining up or waiting for dismissal or at other transition times.

Solving +1 Story Problems

Students use a number strip and cube to solve +1 problems.

Each student needs a number strip (0–10) and a cube.

Pose simple problems for students to solve, for example:

Steve packed 3 snacks for school.
Then he packed 1 more.
How many snacks did Steve pack?

Students place their cube on 3 to start, and then move 1 space to the right to show 1 more.

Continue with other +1 problems.

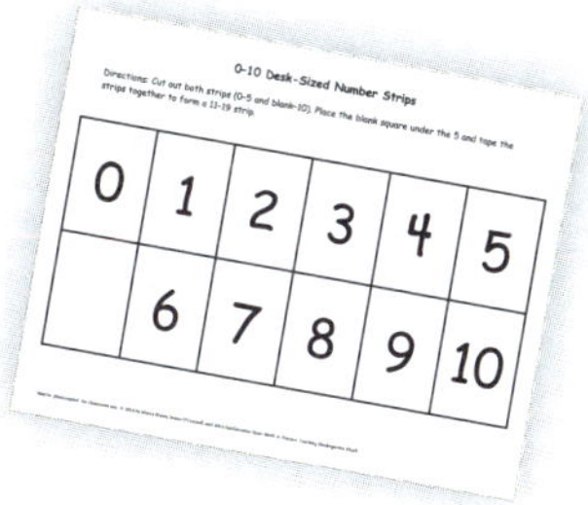

What's 1 More?

Students write the +1 number sentence for simple pictures displayed on the board.

Use an interactive whiteboard to display simple seasonal pictures (e.g., 8 snowmen or 4 bunnies).

Students write the +1 number sentence (e.g., $8 + 1 = 9$ or $4 + 1 = 5$).

Then add 1 more to the picture on the board for students to check their answers.

UNDERSTANDING −1

Students explore subtracting 1, connecting it to the concept of 1 less.

Ten Sly Piranhas

Students are introduced to −1 through the story *Ten Sly Piranhas* by William Wise and use connecting cubes or goldfish crackers to act out subtracting 1.

Before Reading:

Do you know what a piranha is?
What does it mean to be sly?
We are going to read a book about 10 sly piranhas.

Children's literature books that count backward are a great way for students to understand the meaning of −1. As the story counts back, students will learn that −1 is taking 1 away or finding 1 less than the previous number.

The story begins with 10 fish, but listen to see what happens to the number of fish as we read the story.

During Reading:

Read *Ten Sly Piranhas* by William Wise to the class, pausing occasionally to discuss what is happening in the story.

As you read the story, ask:

What is happening to the number of piranhas?

Now there are 5 piranhas, what number do you think will come next?

After Reading:

What did you notice about the numbers in that story?

Retell the story and have students act it out using 10 connecting cubes or goldfish crackers.

How many piranhas were there at the beginning of the story?

What happened?

How did you show that with your cubes?

How many fish are left?

What number sentence could I write to match what happened in the story?

Have students share their ideas.

Write $10 - 1 = 9$ on the board.

What is the 10?

What is the 1?

What is the 9?

Why did I put a minus sign?

Continue with the retelling, having students act it out with the counters and explain the new equations.

Write each new equation directly under the one before it.

$10 - 1 = 9$

$9 - 1 = 8$

$8 - 1 = 7$

After doing several of these problems, begin to ask students to observe the equations and predict how many fish will be left before they show it with their counters.

How many cubes do you think we'll have when we take 1 away? Explain.

After writing all of the −1 equations, ask the students:

Do you notice a pattern when we take away 1?

What are you noticing when you look at what we have already done?

What do you notice about the number sentences we've written?

What if you didn't have counters? How could you take away 1?

Showing -1 on a Ten Frame

Students show numbers on a ten frame and take away 1.

For this small-group lesson, each student needs a ten frame, 10 counters, and a number strip.

Give each student a ten frame, counters, and number strip.

Show the number 6 on the ten frame.

For this lesson, it is helpful if the students show 6 in the traditional arrangement using 5 counters to fill the top row and 1 counter in the bottom left box of the ten frame.

Observe as students build 6 on the frame.

Take away 1 counter.

What number is shown on the frame now?

On the board write 6 − *1* = *5*.

So 6 − 1 equals 5.

What happens when you take away 1 from 6?

Put your finger on the 6 on your number strip.

How many did we have when we took 1 away? Move your finger to that number.

Let's show another number.

Show 8 on your ten frame.

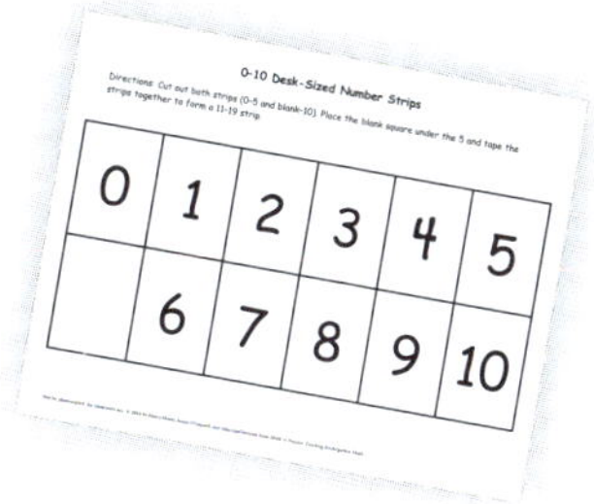

Watch as students place 8 counters.

Take away 1 counter.

What number do you have now?

By touching and moving their fingers on the number strip, students begin to see the action of −1. They recognize the direction they move on the number strip (or in the counting order) to show 1 less.

On the board write *8* − *1* = *7*.

What number is 1 less than 8?

Put your finger on the 8 on your number strip.

How many did we have when we took 1 away? Move your finger to that number.

Let's try another number.

Show 10 on the ten frame.

What number do you think you'll have when you take 1 away? Explain.

Take away 1.

How many counters do you have? Were you right?

Common Error

Students might move their finger to 9 to show 1 more rather than 1 less.

On the board write *10* − *1* = *9*.

Put your finger on the 10 on your number strip.

How many did we have when we took 1 away? Move your finger to that number.

Continue with other numbers. Continue to have students predict what that number − 1 might be before they do it following questioning like:

Predict what 4 − 1 is.

How do you know?

Try it and see if you are right.

Students' ability to correctly predict the difference allows you to see if they have the big idea. Students should be able to explain that 9 is 1 less than 10 or when you take 1 from 10, it is 9.

Have students turn and tell a partner how they know what the answer will be when they subtract 1.

FORMATIVE ASSESSMENT

Show students a card with a −1 expression (e.g., 7 − 1).

Ask students to predict the answer and explain their prediction, then ask them to use the manipulatives to find the answer.

Their predictions help you see if they understand the meaning of −1 and know that it is 1 less or the counting number that comes before it.

Observing the students modeling the expressions with manipulatives helps you see if they are accurately visualizing 7 − 1.

Use the observational notes page to record student information

Additional Children's Literature for -1

These books offer a perfect context for −1 activities.

For each story, students might act out or draw events in the story to show 1 less than the number of pumpkins, fish, monkeys, or bears on each page. They can be asked to write equations to show the −1 facts illustrated in the stories.

Ten Orange Pumpkins by Stephen Savage: Pumpkins disappear in fun ways in this 10–1 rhyming book.

Ten Little Fish by Audrey Wood: The 10 fish swim around to model −1 facts.

Five Little Monkeys Jumping on the Bed or *Five Little Monkeys Sitting in the Tree* by Eileen Christelow: The 5 monkeys fall off the bed or out of the tree to model −1 facts from 5.

Ten Little Bears: A Counting Rhyme by Kathleen Hague: The little bears go throughout their day with rhyming poems to model −1 facts from 10.

DEVELOPING FLUENCY WITH -1

The following activities give students ongoing practice applying their understanding of −1 to become fluent with −1 facts.

Math Racks

Students explore −1 using math racks.

Directions: Math racks are made using cardboard or craft foam, two pipe cleaners, and 20 small beads—10 each of two colors.

1. Cut a piece of craft foam to measure approximately 9 × 4 inches.
2. Use the end of a pipe cleaner to poke a hole near the top left of the foam and push the end through about two inches.

3. On the front, put 5 of the first color beads on the pipe cleaner and then 5 of the other color.
4. Poke another hole straight across near the top right and push the other end of the pipe cleaner through about two inches.
5. Repeat the process near the bottom to form another row of beads.
6. Start this row with 5 of the second color beads and then 5 of the first color.
7. Turn the rack over and twist the top and bottom ends of the pipe cleaners together.
8. Beads are "parked" on the right side and are slid to the left to show a number.

Each student needs a math rack.

Using a math rack, students slide a given number to the left (e.g., 6).

Students can then be asked to show 6 − 1.

Students move 1 bead back to the right leaving the remaining amount to the left (e.g., 6 beads, slide 1 back, 5 beads are remaining).

Students then say and write the math fact: 6 − 1 = 5.

This can be done in small group, with partners, or as a whole-class practice.

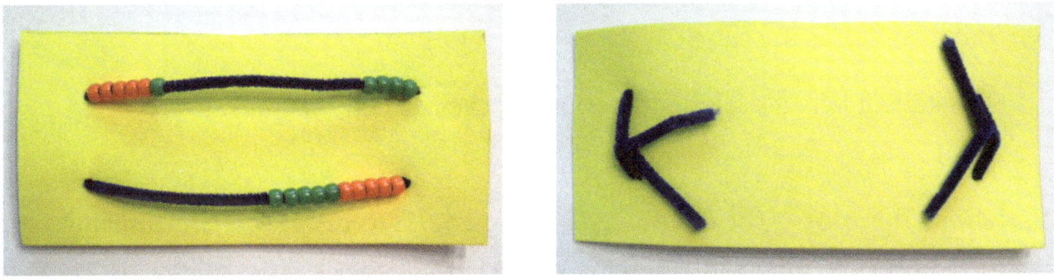

The math rack on the left shows the number 6. Students practice −1 facts by sliding 1 bead back to the right. Students can also practice writing the equation 6 − 1 = 5. The picture on the right shows how the pipe cleaners are twisted on the back of the math rack to secure the beads.

Talk About It/Write About It

How are +1 and −1 alike?

How are +1 and −1 different?

Pop the Balloon

Students draw pictures of balloons to show −1 and then write equations.

The student picks a ten frame card.

He draws that number of balloons on a dry erase board.

Then he crosses out 1 to show it popped.

He writes the equation to show the subtraction story.

Then, he picks another card and repeats.

–1 Partner Pictures

Partners write –1 equations and draw pictures.

Each pair of students needs a 1–10 spinner, paper, and pencil.

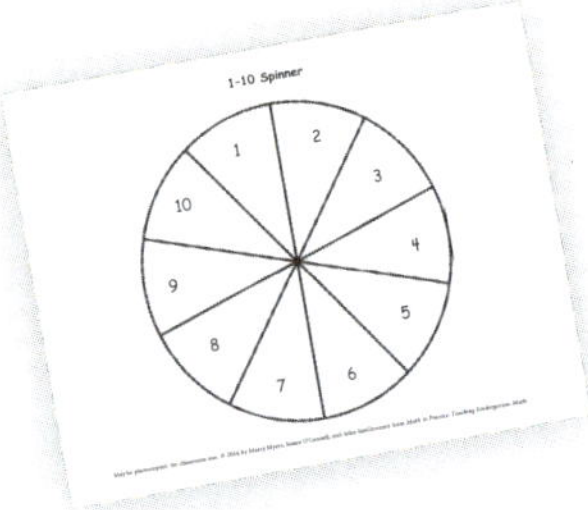

Player A spins a 1–10 spinner and writes a –1 equation for the number spun (e.g., spins 6, so writes $6 - 1 = 5$).

Player B draws a picture to match (e.g., draws 6 peanuts and crosses out 1).

Then players switch roles.

–1 on a Number Strip

In this small-group activity, students show –1 on a number strip.

Each student needs a set of number cards 1–10, a number strip 0–10, and a cube for a marker.

A student picks a number card (1–10).

The student finds that number on the strip and places the cube there to start, and then moves the cube 1 space to the left to show –1.

The student tells the teacher the equation she just demonstrated (e.g., $4 - 1 = 3$).

Variation: This can also be done as a partner game. Player A selects the card and Player B moves the cube and says the math fact. Then, they reverse roles.

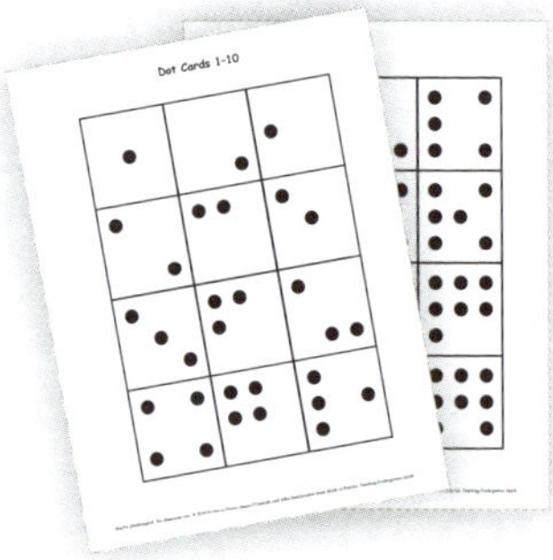

Spin +1 or –1

Using a spinner for +1 or –1, students write equations with a partner.

Prepare a +1/–1 spinner and 0–10 dot cards for each pair.

Player A picks a 1–10 dot card and tells her partner the number of dots on the card (e.g., 5).

Player B spins a +1/–1 spinner and says the math fact for whichever she spun, +1 or –1 (e.g., $5 + 1 = 6$ or $5 - 1 = 4$).

Player A writes the equation.

Switch roles and continue until all of the cards are used.

UNDERSTANDING +0

Students use problems, stories, manipulatives, and discussions to explore adding 0 to discover that it does not add anything to a quantity.

Adding 0

Students explore adding 0 with a problem context and then discuss the equation that shows the story situation.

Each student needs part-part-whole mats and 10 counters for this whole-class lesson.

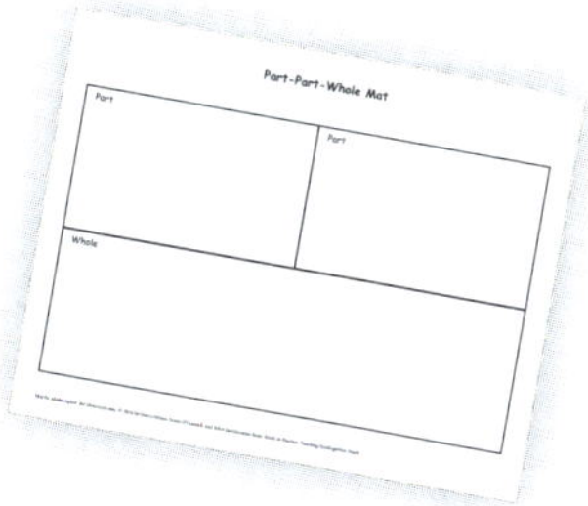

Pose the problem:

> Kellen had 3 apples and Bailey had 2 apples. How many did they have altogether?

Show me on your part-part-whole mat.

Observe as students build it.

How many apples do they have altogether?

How do you know?

If needed, review how to use the part-part-whole mat to show and solve the problem.

Pose the problem:

> Kellen had 3 apples and Bailey didn't have any apples. How many did they have altogether?

Show me on you part-part-whole mat.

Students may be initially confused about having no counters on a part of the mat because previous uses of the mat always had counters on both sides.

Observe as students build it.

How many do they have altogether?

How do you know?

How did you show it on your mat?

Why were there no counters in one part?

How do I show that Bailey has no apples with a number?

Write the expression $3 + 0$ on the board.

What does this show? Why did I write $3 + 0$?

How many apples were there when you put them together?

Does that make sense? Explain.

Complete the equation $3 + 0 = 3$ on the board.

Pose additional problems for students to model on their mats and then write the equations together.

Possible problems:

> Blake had 5 mints and Molly had no mints. How many did they have together?

> There were 4 girls jumping rope. There were no boys jumping rope. How many children were jumping rope?

+0 on Ten Frames

Students have used ten frames throughout kindergarten math. Now they learn that adding 0 adds nothing to the number by showing it on a ten frame. Students write the equation to show +0.

For this small-group lesson, each student needs a set of ten frame cards and dry erase boards to write equations. Also have ten frame mats to model with the group.

Display a large ten frame showing 7.

What number does this show?

If we add 0, how would it change?

What is the total?

How can we write an equation to show that?

Write the equation on your dry erase board.

Observe that students are writing the equation correctly.

Write $7 + 0 = 7$ on the board.

We had 7 and we added 0 to it. So 7 + 0 = 7.

Ask students to choose a ten frame card and write a number sentence to show the number +0.

Observe as they write the equations.

After students have written the number sentence, have some students share with the group.

What number is on your ten frame?

What is ___ + 0?

What number sentence did you write?

Record number sentences on the board as students report.

Did anyone else have the same number sentence?

Let's read some of them.

Have students look at the equations on the board.

$3 + 0 = 3$

$6 + 0 = 6$

$2 + 0 = 2$

$9 + 0 = 9$

What happens to a number when 0 is added?

Does that make sense?

Students can continue to select a new ten frame and write +0 number sentences.

FORMATIVE ASSESSMENT

Show students a number card and have them write the +0 number sentence (e.g., show the number 4, and the student writes $4 + 0 = 4$).

Continue with another number card.

Look to be sure students understand that the sum does not change.

If students have errors with the conventions of writing the equation (e.g., placing the plus signs and equal signs where they belong), show them how to do it correctly, but the assessment here should focus on understanding that adding 0 does not change the quantity. If that is not understood, use manipulatives and have them show you a number, then tell them to add 0 and see what they do. Do they understand that it means not to add anything?

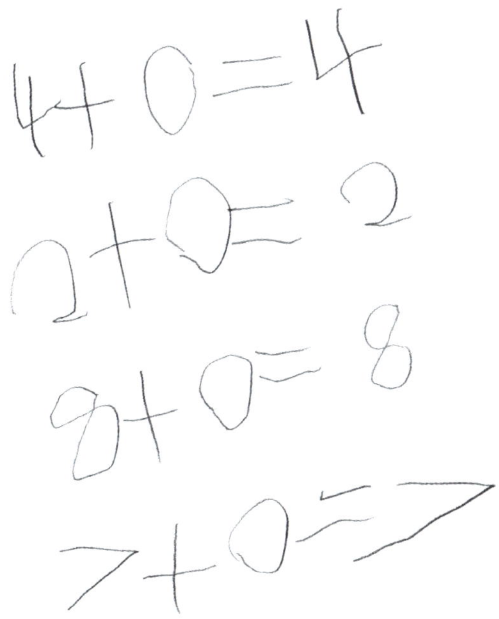

DEVELOPING FLUENCY WITH +0

Students revisit +0 facts to apply their understanding and gain fluency with the facts.

Fact Bracelets

Students use number bracelets and sticky notes to write +0 facts.

Form a circle with a pipe cleaner. Put 1 bead on it and twist the ends of the pipe cleaner together.

Continue with numbers 2–10 using the same color of beads on each bracelet.

It is helpful if the beads are a different color from the pipe cleaner.

To label the number of beads on each bracelet, create a tag by folding a piece of masking tape or adhesive label onto itself around a section of the pipe cleaner.

Label the amount on the tag with a permanent marker

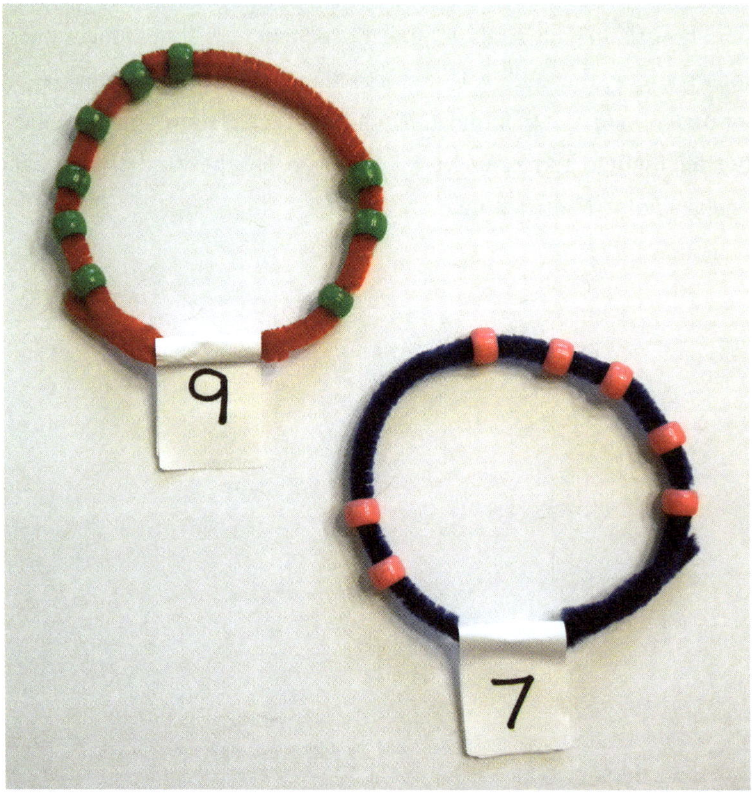

Number bracelets also appear in Module 6.

Sets of bracelets for numbers 1–10 can be stored on ring hooks for each student.

The student selects a number bracelet from the set of 1–10.

Then, she writes the +0 number sentence to match it (e.g., if the student selects the bracelet with 6 beads, she will write *6 + 0 = 6* on a sticky note and place it on or near that bracelet).

Continue with other numbers and students writing the equation on the sticky note to match the bracelet.

+0 Memory

Students play traditional Memory or Concentration to match +0 number sentences to their sums.

Write expressions for +0 on index cards (0 + 0 =, 1 + 0 =, 2 + 0 =, etc.) and write the numbers 0–10 on another set of cards.

Students place the cards facedown on a table.

Students play memory by matching the number sentence to the answer.

If it matches, they keep the cards. If not, they return them to the table.

> **Differentiation**
> *A simpler version of this game is a matching activity in which the number cards 1-10 are lined up, faceup, in front of the partners. Players then take turns picking a +0 fact card from a facedown pile and finding the number card to show the sum until all cards have been matched.*

Counting +0 or +1

Students count sets of items, add 0 or 1, and write the number sentence.

Prepare bags with sets of 1–10 small items in bags.

A student selects a bag and counts the number of items.

Then, the student spins the +0/+1 spinner to either add 0 or 1.

The student writes the +0 or +1 number sentence on his paper.

The student selects another bag of items to count, spins the spinner, and writes the +0/+1 number sentence.

> *Each bag would have a different number of items.*

Dot Cards +0 Partner Game

Students choose a ten frame card or dot card and say and write the corresponding +0 equation.

Each pair of students needs a set of 1–10 dot cards or ten frame cards.

Students place the cards in a stack, facedown, in front of them.

Player A selects a card and states the +0 equation.

Player B writes the equation.

Switch roles and continue with other cards.

Variation: Students roll a 1–6 die and add +0.

UNDERSTANDING -0

Students explore taking 0 from a set of objects, noticing that subtracting 0 does not change the quantity they have.

Thinking Through a Lesson

1, 2, 3 to the Zoo

1, 2, 3 to the Zoo by Eric Carle provides a context for understanding subtracting 0.

Students need cubes or counters to model the problems.

Before reading activities are meant to engage the listeners with the story. The story will then be used as a context to solve math problems.

If students do not know what a zoo is, you may need to develop the context a bit before beginning the story.

Before Reading:

What is your favorite animal at the zoo?

We are going to read a story about a train taking animals to the zoo.

Let's see if any of the animals you named are on the train.

During Reading:

As you read the story, focus students on the counting pattern in the story by asking them to predict the number that will be next.

Students can show their predictions using their fingers as the story is being read. Monitor to see if students are holding up an accurate number of fingers.

After Reading:

Were any of your animals on the train?

Pose the following problem (displaying it on the board or on chart paper):

There were 4 lions on the train.
1 lion got off the train.
How many lions were still on the train?

SMP4

How could we show this problem? (Act it out. *Or*, Use counters. *Or*, Draw a picture.)

Beginning with a familiar −1 problem allows students to revisit that concept and allows for a transition to −0.

Give students some cubes.

Show me the lions with your cubes.

Reread the first line of the problem: There were 4 lions on the train.

How can you show that? (Use 4 cubes.)

What are those cubes supposed to be? (The lions.)

What happened next in the story problem?

Reread the problem.

SMP1

Tell your partner what happened. (A lion got off the train.)

How can you show this? (Take 1 cube away.)

How many lions were still on the train? (3.)

How do you know? (I counted them. *Or*, I knew that 1 less than 4 is 3.)

Models and teacher questioning remind students about the subtraction structure of taking from.

Write *4 − 1 = 3* on the board.

4 lions minus 1 lion leaves 3 lions on the train.

Why is this subtraction? (We took 1 away.)

Have students put their counters back.

That was a silly problem. Do you think the lions would get off the train before it got to the zoo?

I think they might stay on the train.

Pose the following problem:

> There are 4 lions on the train.
>
> 0 lions get off the train.
>
> How many lions are still on the train?

Show me the lions with your cubes.

How can you show that? (Use 4 cubes.)

What happened next in the story problem?

Reread the problem.

Did any lions leave the train? (No.)

How can you show this?

Students may be unsure.

What does it mean that 0 lions leave the train? (None leave. *Or*, 0 means none.)

Can you take 0 lions from your group? (You don't take any cubes because there are no lions left.)

How many lions are still on the train? (4.)

Does that make sense? There were 4 on the train at the beginning of the story problem. (Yes, because it is the same, none left.)

Write *4 − 0 = 4* on the board.

4 lions minus 0 lions leaves 4 lions still on the train.

Students should be consistently answering these questions with turn-and-talk strategies rather than one student answering the question. All students need an opportunity to think about and express their methods and answers.

Posing the same problem with different data allows students to explore −0 in a context they have already seen.

When students have limited literacy skills, it is important to reread the problem several times throughout the lesson to remind students of the story events and data. While you are reading the same words, they are focusing on different aspects of the problem-solving task (e.g., understanding the story events, listening for the data, checking to see if their answer makes sense).

Explorations with hands-on materials allow students to make sense of the idea of taking 0 from a quantity.

By seeing the way you write the equation and hearing your summary of what it says, students are beginning to develop an understanding of what the numbers and symbols in the equation represent.

Pose additional problems focused on the zoo picture on the last page of the book.

Pose the problem:

> There are 10 birds at the zoo.
> 0 birds flew away.
> How many birds are at the zoo?

Students are now asked to interpret the numbers in the equation. Listen to see which students are making connections between the story events and the numbers in the equation.

Have students use their cubes to show the problem.

Have them share their answer as well as their thinking about how they solved it.

Write *10 − 0 = 10.*

> *Turn and tell your partner what each number in my number sentence shows.*

Listen as students share their ideas with partners.

Students may not immediately be able to explain each number in the equation, but consistently asking them to connect these numbers to the story situation helps to build that skill.

Point to the first 10 in the equation.

> *What is this 10?* (The birds at the start.)
> *What is the 0?* (How many flew away.)

Listen for students to say that subtracting 0 means that nothing is taken away or that subtracting 0 means you still have the same amount you started with.

Point to the 10 showing the difference.

> *What does this 10 show?* (How many were still there.)
> *10 birds minus 0 birds leaves 10 birds.*

Ask students to turn and tell their partners what it means to subtract 0.

Continue with other problems about the zoo as time permits.

> 7 seals are in the pool.
> They love to swim.
> 0 seals get out of the pool.
> How many seals are in the pool now?

Differentiation

Some students may need additional help thinking about removing 0 from a group. Additional opportunities to act out problems or use manipulatives to show −0 situations may be indicated for some, and others may be able to move to drawings to show the story situations.

This lesson focuses on understanding −0 using concrete materials. A similar lesson using drawings could extend students' understanding. As students draw pictures of −0 problems, they realize they do not cross off any items on their pictures.

FORMATIVE ASSESSMENT

Show students 7 counters.

> *What happens when you take away 0?*

Students should be able to explain that nothing is taken away because 0 is nothing.

Record notes of students' responses.

Use the observational notes page.

If a student is not able to explain that there are still 7 counters because subtracting 0 does not change the number, they will need additional practice using manipulatives and exploring the meaning of minus 0.

Writing -0

In this small-group activity, students practice writing –0 equations.

Write the numbers 1–10 on cards and have dry erase boards and markers for each student.

I am going to show you a number.

You could think of these numbers as representing pennies in your piggy bank.

I want you to think about how we would show –0.

What if you didn't take any pennies out of your bank?

Show the 5 card.

You have 5 pennies in your piggy bank.

Draw me your pennies.

Think about taking 0 pennies out of your bank. How many are still in your bank?

What –0 number sentence can you write?

Show me on your board.

How do you know you are correct? Explain your number sentence.

If you took no pennies out of your bank, how many would be left?

Have students erase their boards.

Show students the 9 card.

Now you have 9 pennies in your bank.

Draw the 9 pennies. Think about taking 0 pennies out of the bank. How many are still in the bank?

What –0 number sentence can you write?

Allow students time to write, then have them share their responses with the group.

Show me your number sentence.

What does the number sentence tell us?

If you took no pennies out of your bank, how many would be left?

Continue with other number cards.

Differentiation

Some students may be able to work with numbers 10–20 to practice -0.

Although the focus of this lesson is on writing equations, the context of pennies in a piggy bank makes it less abstract and continues to remind students of the meaning of the numbers in the equations.

DEVELOPING FLUENCY WITH -0

Through interactive activities, students revisit –0 facts and continue to apply their understanding that the quantity stays the same when subtracting 0.

Vocabulary

add
equal
math fact
minus
one less than
one more than
plus
same as
subtract

ONLINE RESOURCES

General Resources

Additional Problems
Dot Cards 0–10
Number Cards 0–10
Number Strip
Part-Part-Whole Mat
Ten Frame Cards
Ten Frame Mats
Understanding Facts Student Observation Notes

Resources for Specific Activities

+0/+1 Spinner
+1/−1 Spinner
1–10 Spinner

These resources are available at http://hein.pub/MathinPractice, keycode MIPGK.

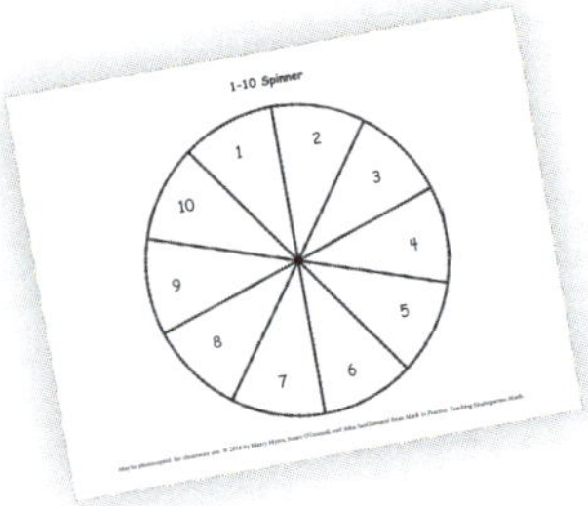

Hopscotch

Students play hopscotch and say −0 equations.

Create a hopscotch board on the floor with tape or draw one outside on the playground.

Students toss a beanbag and hop to that number.

Students say the −0 equation to pick up the beanbag and hop back.

Ten Frame Subtraction

Students show −0 on ten frames and write equations.

Place 1–10 number cards, counters, a ten frame, paper, and pencil at the center.

Students work with partners to pick a 1–10 number card and use counters to show that number on a ten frame.

They say the −0 number sentence for the number modeled.

Then they write the equation on a piece of paper.

Continue until all cards have been used.

Talk About It/Write About It

How are +0 and −0 alike?

What does it mean to subtract 0?

Tell a story problem that goes with 5 − 0.

Partner Towers

Students make towers and write equations with partners.

Each pair needs 10 connecting cubes, 1–10 cards, and a dry erase board with marker.

Player A picks a card and creates a tower with that number of connecting cubes.

Player B writes the −0 equation to match on the dry erase board.

Player A checks and gives a thumbs up or helps to rewrite it.

Then players switch jobs.

Spin and Draw −0

Each student needs a 1–10 spinner, paper, and pencil.

Students spin a 1–10 spinner and draw a picture to show that number.

Students then write the −0 equation to match their picture.

MODULE 10

Exploring Measurement

About the Math K.MD.1; K.MD.2

Kindergarten students are introduced to attributes of objects that can be measured. They explore height, length, capacity (how much liquid something holds), and weight as measurable.

The key ideas focused on in this module include:

- **exploring measurable attributes of objects**
- **describing measurable attributes with words like *short, tall, long, heavy, light***
- **comparing two objects by a measurable attribute.**

Our students learn about measurable attributes through investigations, discussions, and examples. Students talk about *tall* or *short* and observe that some flowers in the school garden are tall and others are short. They pick up objects and identify that one is heavy and another is light. They fill containers and notice that one container holds more water than another. Students sort objects as heavy or light, or search for classroom objects that are long or short.

As our students observe and handle items, we ask them to describe what they see or feel. Using words like *taller*, *shorter*, *lighter*, *heavier*, *holds more liquid*, *wider*, or *fatter*, they are able to tell us about the objects. Creating Math Talk charts or word walls, and illustrating the words with pictures that give clues to their meanings, helps our students gain familiarity with these words (see Figure 10.1). Journal writing allows students to share their ideas as they label one page *heavy* and the other *light*, or one page *short* and the other *tall*, and then draw pictures of objects that fit their measurement labels.

Figure 10.1 Math Talk charts, or word walls, can be illustrated with pictures or with students' own drawings that show the meaning of each word.

Comparing two objects by a specific attribute extends students' measurement understanding. Holding two objects, students decide which is heavier or lighter than the other. They observe two classmates to see who is taller or shorter. Students might complete journal entries, with words or drawings, that focus on comparison.

___ is shorter than ___.

___ is taller than ___.

___ is heavier than ___.

___ is lighter than ___.

___ holds more water than ___.

___ holds less water than ___.

___ is longer than ___.

As they declare that the can of peaches is bigger than the pencil, we might challenge them for more specific information: Is it wider or fatter? Is it taller?

As students explore length or height, misunderstandings surface. What happens when we don't align the endpoints when we are trying to figure out which item is longer or taller (see Figure 10.2)? And if a long piece of string is curved and appears shorter than another piece of string, as in Figure 10.3, is it really shorter (conservation of measurement)? As students explore measurement with real objects, questions about what they see lead to big ideas that deepen their understanding.

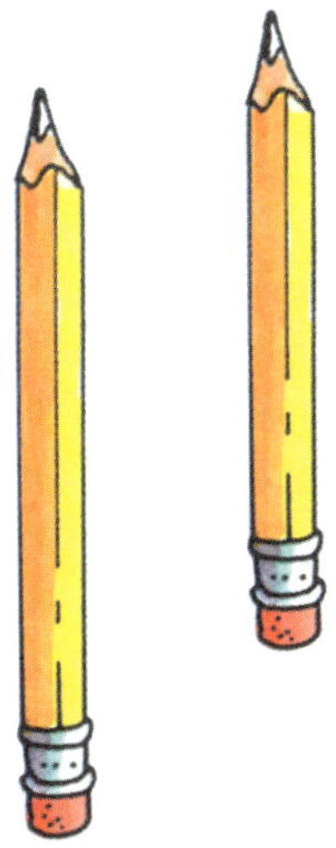

Figure 10.2 Which pencil is taller? One may look taller, but does it matter that they don't start at the same place (the endpoints are not aligned)?

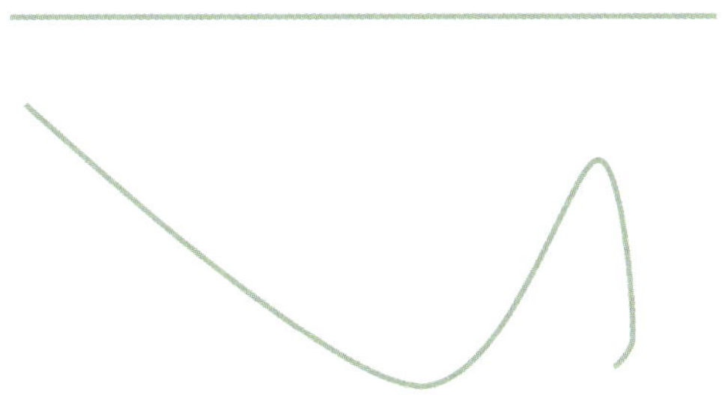

Figure 10.3 Which string is longer?

Introducing a balance scale allows students to make comparisons about weight. Predicting which object will be heavier and then testing to see if their predictions are correct fine-tunes their understanding of weight.

Our students are engaged when they are asked to answer questions or solve problems related to measurement. Posing problems like the following leads to observations, discussions, and insights:

- What objects can you find that are longer than the pencil?
- What objects can you find that are shorter than the pencil?
- Which container holds more water than the sippy cup?
- Which container holds less water than the sippy cup?
- Which items are heavier than this book?
- Which items are lighter than this book?

And posing problems in which one item is bigger, but lighter, than another (e.g., a Wiffle® ball and a golf ball) shows our students that size and weight are not always related.

Learning Goals

I can tell which object is shorter or taller/longer.

I can tell which object is heavier or lighter.

I can tell which container holds more or less.

Lessons in This Module

Describing Length and Height

Long, Tall, Short

Is It Longer? Is It Shorter?

Aligning Endpoints

Using Yarn Snakes to Compare Lengths

Describing Weight

Comparing the Weight of Books

Exploring Heavy and Light

Just a Little Bit

Describing Capacity

How Much Will It Hold?

Chores

Goldilocks and the Three Bears

Exploring the Progression

PREVIOUS	NOW	NEXT
Preschool	**Grade K**	**Grade 1**
Possible familiarity with the words *heavy, light, long, tall, short* but they may not have been used to compare items	Exploring measurable attributes; describing measurable attributes with words like *heavy, light, long, tall, short*; comparing two objects by a measurable attribute	Measuring items by laying a shorter object end to end; ordering three objects by length

Ideas for Instruction and Assessment

DESCRIBING LENGTH AND HEIGHT

Students explore length and height with hands-on investigations and make discoveries about aligning endpoints and considering whether lines are curved or straight.

Long, Tall, Short

Students learn to use measurement vocabulary to describe the length and height of objects.

Materials: a collection of items that can be measured by height or length (e.g., a short tower and a tall tower, a short piece of yarn and a longer piece of yarn), index cards that say *tall*, *short*, and *long* for labeling the objects

Show a short and a tall tower, both made of the same color connecting cubes.

How could you describe these towers to me?

Students may describe the towers by color (red) or as big and small. This can be the beginning of the discussion about measurable attributes.

Carlos says the towers are red. Do you agree?

Roxy says the towers are big and small. Which is big and which is small?

Have students point to the big tower and the small tower.

Can you think of other words for big *and* small*?* (*Tall* and *short*.)

Which tower is tall?

Have students point to the tall tower.

Which tower is short?

Have students point to the short tower.

We are describing the height of the towers.

Write *short* and *tall* on the Math Talk chart and draw a picture of the short and tall towers next to each word.

Label the towers as short and tall with the index cards.

What are some other things in our classroom that are short and tall?

We can use the words short *and* tall *to describe items when we talk about their height.*

Continue with the short and long yarn.

What words can we use to describe these pieces of yarn?

Students may describe the yarn by color or as tall and short.

Jackson noticed that one piece of yarn is short and the other is tall.

Which piece of yarn is short?

Have students point to the short piece of yarn.

With our towers, we saw one was tall. It went higher than the other tower when we placed them both on the table.

This yarn is long. It goes farther when we lay them both on the table.

We use the words short *and* long *when we talk about length.*

Add the words *short* and *long* to the Math Talk chart, along with quick illustrations.

Label the yarn as long and short with the index cards.

What are some things in the school that are long or short?

Students may initially notice the color of the towers. Although the color does describe the towers, it doesn't describe them in a measurable way. Color is a nonmeasurable attribute. Our goal is to introduce students to measurable ways to describe the towers.

As you continue with measurement, use short, tall, and long, to describe items throughout the classroom and school.

Thinking Through a Lesson

Is It Longer? Is It Shorter?

Students count and build a train of 10 connecting cubes, then search for objects in the room that are longer or shorter than their train of cubes.

Materials: 10 connecting cubes for each student, 1 set of vocabulary cards for *longer* and *shorter* for the teacher to use

*We described the length of some pieces of yarn. Which words do we use to describe length? (*Long *and* short*.)*

Make a train of 4 cubes.

Can you make a train that is longer than 4 cubes?

Watch as students make trains to see if they understand what it means to be longer.

Hold up the train that is longer than the other one.

What does it mean to be longer? (It is bigger. *Or,* It has more cubes so it is longer when you lay them down next to each other.)

Kindergarten students often rely on the terms big and small/little to describe the length of objects. Although these terms are not inaccurate, explain that length can be described best using terms such as tall, long, and short.

SMP4

Asking students to create trains that are longer or shorter allows you to assess their understanding of the concept.

SMP6

Show the vocabulary card that says *longer*.

Counting practice is integrated into this task.

Have students take apart their trains.

Make a train of 6 cubes.

Can you make a train that is shorter than 6 cubes?

Showing the vocabulary card reminds students of the word and helps them become familiar with the look of the word.

Watch as students make trains to see if they understand what it means to be shorter.

Hold up the train that is shorter than the other one.

What does it mean to be shorter? (It is smaller. *Or*, It has less cubes so it is shorter when you lay them down next to each other.)

Show the vocabulary card that says *shorter*.

Have students take apart their trains.

Differentiation

Some students may benefit from working with partners to hear their partner's thinking as they search for an item that is longer.

We are going to use these connecting cubes to help us find objects in our classroom that are longer or shorter than our cubes.

Make a train of 10 cubes.

Can you find something in our classroom that is longer than the 10 cubes?

Walk around the room using your train of cubes to find something that is longer.

When you find something, bring it with you back to the group.

Allow students time to search in the classroom.

Let's see what you found that is longer than your train of 10 cubes.

Turn and talk with your neighbor about what you found.

When you answer, make sure you use the word longer.

How can you prove to your partner that your object is longer than your cube train? (Put them next to each other.)

Having students turn and share gets everyone talking and allows you to listen in on their conversations.

Practice using vocabulary reinforces students' understanding of the words.

SMP3, SMP5

Have students share what they found with partners, telling that it is longer and showing how they know (by putting their train of 10 cubes next to it).

Then, have a few share with the group.

Now that you have found something longer than the train, we need to find objects that are shorter than the train.

Keep your longer object at your spot while you find something that is shorter than the train.

As students share, they can align their object with the cube train to prove that it is longer. They are using the cube train as a tool to compare.

Allow students time to search in the classroom.

When students return to the group, have them turn and talk to their neighbors telling them what they found that was shorter and proving that it is shorter. Remind them to use the word *shorter* as they talk about the objects.

As students are responding, take observational notes: Who shows understanding of longer and shorter?

Have a couple students share their examples with the group.

Next, students compare the two objects they found in the classroom.

Let's look at the two objects that you found.
Put your two objects on the carpet.
Touch the longer object.
Check your neighbor to see if he or she is touching the longer object.
Touch the shorter object.
Check your neighbor to see if he or she is touching the shorter object.

Have them use the word *longer* to tell their neighbor which object is longer.

Have them use the word *shorter* to tell their neighbor which object is shorter.

Have them draw their objects on the recording sheet.

Aligning Endpoints

Students learn to align endpoints when they compare the lengths of items.

Materials: two pencils—one long and one short, preferably each a different color; document camera (optional but helpful)

Hold each pencil in a hand but have the shorter pencil appear to be longer by sticking it up higher in your hand.

Which pencil is longer?
Let's check which is longer.

Then put the pencils on a document camera, if available, in different directions and not aligned.

Which pencil is longer?
How could I figure it out?
What if I put them this way?

Place the pencils side by side and line up the flat endpoints.

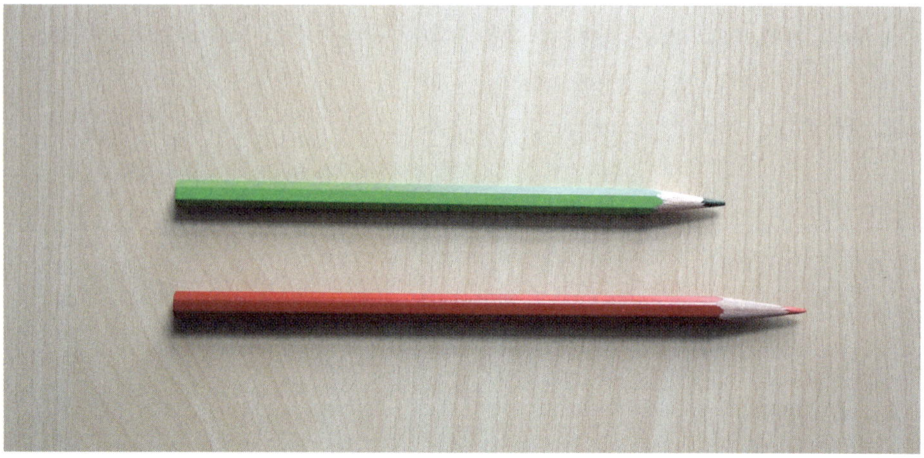

Now can we tell which pencil is longer?

How did that help you?

What did I do?

Put them next to each other again, but this time do not align the endpoints and make the shorter pencil appear longer.

Which pencil looks longer now?

Is it really longer?

What can I do to show how long they really are?

Align the endpoints as students watch.

What did I do?

Now, can you see which is longer?

Why is it important that we line up where they start, the endpoints?

Repeat on the document camera with two other pencils of similar, but slightly different, length.

Have the shorter pencil displayed vertically and the longer pencil displayed horizontally.

Which pencil is longer?

How can we figure out which pencil is longer?

Reposition the pencils to align the endpoints.

Which is longer?

Continue with other examples.

Other examples include pieces of yarn, various markers, or strips of paper.

Then, have students compare two items by lining them up.

Watch to see if they align the endpoints.

Have them tell you which is longer and prove it to you by showing their pencils lined up.

Using Yarn Snakes to Compare Lengths

Students explore conservation of measurement by comparing the lengths of yarn snakes.

Materials: enough pieces of yarn (or ribbon) in various lengths for each student to have two pieces

Lay two pieces of yarn on the table. The endpoints should not be aligned.

Let's pretend these are snakes. How can we compare the lengths of these snakes?

Align the endpoints and straighten the pieces of yarn.

What words can we use to compare the lengths?

Turn and talk with your partner.

Talk about which snake is long and which is short.

Listen as students talk about the length of the snakes.

How did you describe the length of the snakes?

Can you use the word longer *to talk about the snakes?*

Can you use the word shorter *to talk about the snakes?*

It's your turn to compare lengths.

Choose two snakes and be ready to tell the group about the snakes using long *and* short *or* longer *and* shorter.

Have students report about the lengths of their yarn.

Students do not have to measure, just compare lengths.

How can you use the words long *and* short *to describe the snakes?*

How can you use the words longer *and* shorter *to describe the snakes?*

After students have shared their examples, show two more pieces of yarn, this time curving one to appear shorter than the other.

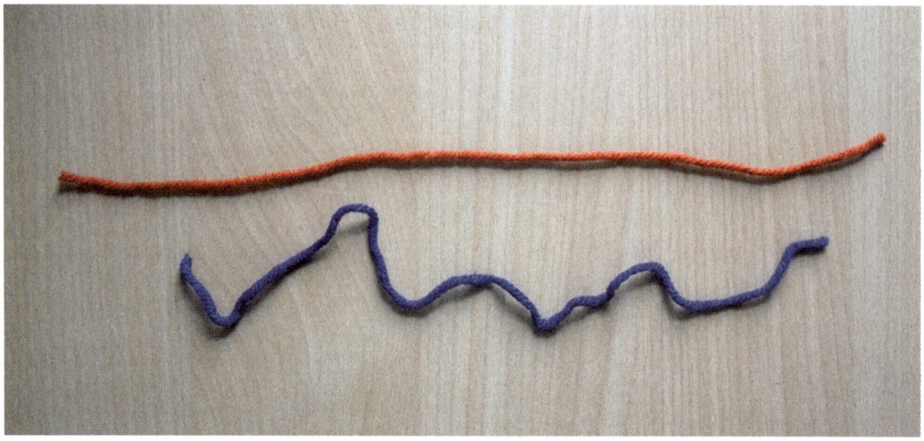

Look at these two snakes.

Which is longer?

Common Misconception

Students may think that the orange piece is longer because the blue piece is curved

Students may have different opinions because the blue snake appears shorter.

Who thinks the orange is longer? Why?

Who thinks the blue is longer? Why?

How could we find out which is longer?

Have two students stretch out the pieces of yarn. Talk about the endpoints and whether they are at the same starting point.

Which is longer?

Try it with your snakes. Can you make the longer snake look shorter?

Differentiation

Some students may need additional support with understanding that the curved yarn is longer than the shorter straight piece of yarn.

Allow students time to work.

Does it look shorter?

Is it still the longer snake?

Take a look at your neighbor's. How did they curve their longer piece? Does it look shorter? Stretch it out to see that it is really the longer piece of string.

Continue using other pieces of yarn and comparing straight and curved lengths.

FORMATIVE ASSESSMENT

Put two pieces of string on the table. Have the longer one curved so it looks shorter.

Ask the student to tell you which is longer.

They should straighten the curved string and align the endpoints of the strings to compare them.

If students look at the strings and determine the curved one is shorter without moving it and realigning the endpoints, ask them what it means to be longer. Ask them how they might be able to figure out which is longer. If they are unable to answer, stretch the longer string out and ask them which is longer. Make notes about their responses and continue lessons related to the measurement concepts.

Use the observational notes page.

Additional Ideas for Support and Practice

The following ideas extend students' understanding of comparing length using *long* and *short* and provide meaningful practice.

Longer or Shorter Than an Index Card

Students draw pictures or label items in the classroom that are longer and shorter than an index card.

Give each student an index card.

Explain that on one side of the card they will draw/label a picture of something in the classroom that is longer than the card.

On the other side they will draw/label a picture of something that is shorter than the card.

Give students time to explore the classroom using their index card to record their findings.

Once everyone has completed the task, have students share what they found.

Comparing Towers

Students build towers of connecting cubes and compare lengths with a partner.

Materials: bags of connecting cubes with a different number of cubes in each bag

Students will connect the cubes in their bag to make a tower.

Partners compare their towers.

Whose tower is taller? Whose is shorter?

Then students can switch partners to compare towers again.

Repeat by having students put the cubes back in the bags and switch bags.

If partners have towers that are the same number of cubes, share with the class that the towers are the same length.

The Best Bug Parade

Students learn to describe bugs as long, short, big, and small as the bugs are on parade.

Before Reading:

Ask students to name a bug that is very small, like a ladybug.

Then ask students to name a bug that is very big like a beetle or long like a walking stick.

Tell them that you will be reading a story about bugs that are different sizes.

During Reading:

Read *The Best Bug Parade* by Stuart J. Murphy.

As you read, discuss the size of the bugs and the words used to describe them.

After Reading:

Talk about the bugs that were long and big and short and small.

Then write about the size of the bugs.

Fold a piece of construction paper in half.

Students draw a picture of a long bug in the top part and label it as long.

Students draw a picture of a short bug in the bottom part and label it as short.

Sidewalk Chalk

Students draw and label long and short lines on the playground with chalk.

Give each student a piece of chalk.

Students draw a line that is long and label it.

Then, they draw a line that is short and label it.

When everyone has drawn a set of lines, have students walk around to look at other sets of lines.

Stand and freeze on a short line.

Have students continue to walk around.

Stand and freeze on a long line.

Repeat.

Chain Links

Students use plastic chain links to create long and short chains.

Write directions for long and short chains on cards to place at the center with multicolor chain links:

- long green chain; short red chain
- long blue chain; short yellow chain
- long red chain; short blue chain.

Students create long and short chains for each card and align the endpoints to compare the lengths.

This student makes a short yellow chain and a long green chain. He aligns the endpoints to compare the lengths of the chains.

Talk About It/Write About It

Draw a flower and color it red. Draw a shorter flower and color it blue.

Draw two trees. Circle the shorter one.

DESCRIBING WEIGHT

Students explore weight using pan balances and describe items as heavy and light.

Comparing the Weight of Books

In this small-group lesson, students learn to use measurement vocabulary to compare the weight of objects.

Materials: two books (one heavy, one light), vocabulary cards for *heavy* and *light*

Show the students two books: one very heavy book like a dictionary and one very light book like a paperback.

What words could you use to tell about or compare these books?

Students may describe the books as big and small or short and long.

Students may notice that one book has pictures and the other does not or that the books are different colors. These are nonmeasurable attributes. Our focus will be on helping them identify measurable attributes of the books.

Place the books side by side to connect to previous discussions about length. Have students tell you which book is shorter or longer.

Let's have Ryan hold these books.

Put one book in your right hand and the other book in your left hand.

Ryan, what do you notice?

Why is it hard to hold this big book in one hand?

Pass each book around the group so each student can hold it.

If this book is heavy, how can we describe the other book?

We use the words heavy *and* light *to describe the weight of items.*

Label the books with the cards heavy and light.

Turn to your partner and name some items in our classroom that are heavy.

Would they be easy or hard to pick up?

Tell your partner some things in our classroom that are light.

Would they be easy or hard to pick up?

Tell your partner a heavy animal.

Tell your partner a light animal.

We use the words heavy *and* light *to label items when we talk about their weight.*

As you continue with measurement, use heavy and light to describe items throughout the classroom and school.

Write the words *heavy* and *light* on the Math Talk chart. Include a picture by each word of one of the items the class described as heavy or light.

Have students fold a paper in half. On one side, have them write *heavy* and on the other side, have them write *light*.

Have them draw things that belong in each section, then have them share their drawings and examples of heavy and light with the class.

Exploring Heavy and Light

Students work with a partner to use a balance and manipulatives to compare weights of objects.

If this is the first time students have used pan balances, provide some instruction on the proper use. Also, give students some exploration time where they can freely place objects on the balance before instruction begins.

Materials: several balances, a collection of items or manipulatives to weigh, vocabulary cards for *heavy* and *light*

When you compare the weight of two books, how could we describe the one that is hard to hold?

Refer to the *heavy* vocabulary card.

What would we call something that is easy to hold, that is not heavy?

Refer to the *light* vocabulary card.

We have a special tool to help us compare heavy and light. This is called a balance.

Give each group or set of partners a balance to share. Give them a large wooden cube (e.g., from a set of solid shapes) and some hexagon pattern blocks.

Put one large wooden cube on one side and three yellow hexagons on the other side.

What happened to the balance?

Why do you think the side with the cube went down? (The cube is heavier than three hexagons.)

Can someone use the word lighter *to compare the hexagons to the cube?*

Look at the balance. What happened to the side that is lighter?

Students can pick up the cube and hexagons to compare the weight.

Have students hold the wooden cube in one hand and the three hexagons in the other to feel the heavier and lighter objects.

Now it's your turn to choose some objects to weigh.

Have various manipulatives available for the students to select.

We have lots of manipulatives here for you to use.

Work with your partner to put objects on one side of the balance.

See which side is heavy and which side is light.

Be ready to share with the group which side was heavy and which was light.

After students have had time to explore the balance, have them share with the group.

Students need to use heavy and light as they describe the balances.

What did you find that was heavy?

What did you find that was light?

Can you tell me what was heavier than something or what was lighter?

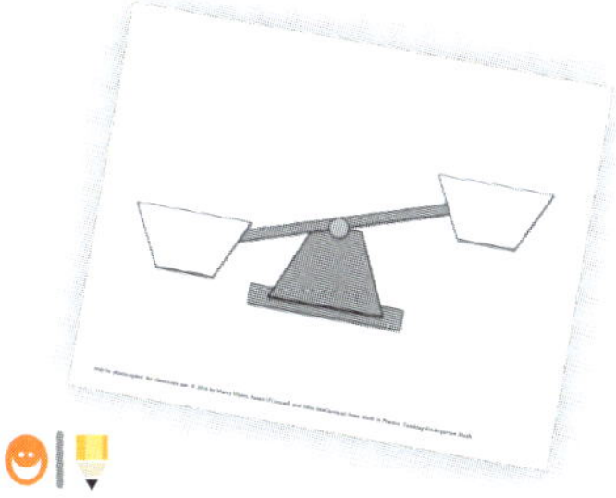

Students can continue with other manipulatives to compare weights and describe them using *heavier* or *lighter*.

After working with manipulatives, students can draw pictures to show what they found to be heavy and light using the balance.

Just a Little Bit

Students use a balance to reenact the story and to show heavy and light.

Before Reading:

Show the students the picture of the seesaw in the book and a classroom balance.

Have you ever played on a seesaw?

How are the seesaw and balance alike?

What do you know about the weight of an elephant?

During Reading:

Read *Just a Little Bit* by Ann Tompert.

As you begin reading, ask students how many animals they think it will take to move the elephant on the seesaw.

After Reading:

Review that it took ten animals to move the heavy elephant: ten smaller animals are heavier than one elephant.

Give groups of students a balance to use.

Put an object in one side of the balance to be the heavy elephant.

How many smaller cubes do you think it will take to balance the heavy cube?

Add some smaller cubes to the other side to be the animals on the seesaw.

Count the cubes as you add them.

How many did it take to balance our elephant?

Students record their results on the Balance recording sheet by drawing the one larger cube and the number of smaller cubes.

Common Error

Observe closely as students are placing their smaller cubes on the scale to make sure they are counting accurately. Students often become distracted by the scale and miscount.

FORMATIVE ASSESSMENT

Option 1

Give students two objects to hold.

Ask them to tell you which is heavier and which is lighter.

Record observational notes of student work and responses.

Use the observational notes page.

Students should be able to tell you the heavier and lighter object using the correct vocabulary.

If they do not know which is heavier, they may have a lack of understanding of the concept of heavier and need more work comparing the weights of concrete objects. Can they tell what it means to be heavy or name things that are heavy?

If they are confusing the terms *heavier* and *lighter*, they need more opportunities to describe concrete objects using the terms.

Option 2

Assess students as they work with a balance.

Place a variety of manipulatives and items on the table in front of the students.

Use the balance to show heavy and light with these items.

Observe as students work to place items in the balance.

The expectation is that students understand the concepts of heavy and light and can describe objects using those terms as well as heavier and lighter. We are not assessing whether they can use a balance but whether they understand the concepts of heavy and light.

Have them tell you about the items using the words *heavier* and *lighter*.

Have them explain how they know an item is heavier or lighter.

Record observational notes of student work and responses.

Additional Ideas for Support and Practice

The following ideas extend students' understanding of heavy and light and provide meaningful practice.

Calculating How Heavy

Using a manipulative of choice, partners balance a block by adding manipulatives to the other side of the balance.

Working in partners, Player A places a square building block in the double pan balance.

Using a manipulative of choice, Player B balances the block by adding manipulatives to the other side of the balance.

Once the two sides are balanced, the students will count and talk about how many manipulatives were needed.

Note the connection to counting skills from Modules 1–4.

Provide a sentence frame for students to record the results in their math journals.

__________ balances 1 block (e.g., 8 bears balances 1 block or 15 spiders balances 1 block).

Repeat with different manipulatives.

Balancing Act

Students listen to a story about animals balancing on a teeter-totter to discuss heavy and light.

Differentiation

For students who struggle with understanding the idea of balancing to compare weight, this story revisits the concept and gives another brief look at the basic understanding of identifying what it means to be balanced and what it looks like when one side is heavier.

Before Reading:

Have you ever been on a teeter-totter at a playground?

You may have to show a picture of one or describe one because some playgrounds no longer have them.

We are going to read a story about some animals who try to balance on a teeter-totter they made out of a rock and a stick.

During Reading:

Read *Balancing Act* by Ellen Stoll Walsh.

Pause when the mice are balanced to talk about what it means (they are at the same level, one is not higher or lower than the other).

After Reading:

When was the teeter-totter balanced?
Why was the teeter-totter not balanced when the salamander got on?
After they got balanced again, what happened?

Do a picture walk-through of the story and have students tell you (with a thumbs up) when the teeter-totter was balanced and when it was not (with a thumbs down).

When it was not balanced, have them tell you which side was heavier and how they knew.

Bring out a balance scale and allow students to try to balance it with weights or objects.

When done, have students draw a picture of the teeter-totter from the story when it was balanced.

Which Is Heavier (or Lighter)?

Students use a balance scale to determine which is heavier.

In a small group, give pairs of students a balance and sets of manipulatives to determine which is heavier.

They use a balance scale to determine which is heavier, 10 Unifix® cubes or 10 trapezoid pattern blocks.

When students have identified which group is heavier, ask them which group is lighter.

Continue with other sets:

6 counters or 6 crayons?
5 counters or 5 connecting cubes?
10 craft sticks or 10 dominoes?

The numbers are to give additional counting practice as students explore measurement ideas.

This student uses a balance to compare teddy bear counters. She is counting to find out which is heavier: 5 large teddy bears or 5 small teddy bears.

A Pig Is Big

Students gain an understanding of measurement vocabulary as they read about items that get progressively bigger.

Before Reading:

What is bigger than a pig?
What is heavier than a pig?
What is taller than a pig?

During Reading:

Read *A Pig Is Big* by Douglas Florian.

As you read, discuss the size of the items in the story and use measurement vocabulary to compare them.

After Reading:

Practice measurement vocabulary to build comparing skills.

What is longer than a pencil?
What is shorter than a crayon?
What is heavier than this block?
What is lighter than this book?

As a class, create books or charts that use the measurement vocabulary to demonstrate the words.

My Shoe

Students practice using *heavy* and *light* with a partner.

Player A takes off a shoe and places it on one side of the balance.

Then both partners find one object that is heavier and one object that is lighter than the shoe.

They test their predictions with Player A's shoe.

Then Player B takes off a shoe and places it on one side of the balance.

Players use the same items to compare Player B's shoe.

DESCRIBING CAPACITY

Students explore capacity through interactive experiences as they fill cups and containers to see which holds more or less.

How Much Will It Hold?

Ounces are used to identify approximate sizes for the cups used in the lesson. Sizes of the containers do not have to be exact and the specific measurements are not shared with students. Kindergarten students do not need to know ounces as a measurement unit.

Students describe how much an object will hold by using measurement vocabulary.

Materials: three cups of different sizes: approximately 32 ounces (large), 8 ounces (medium), and 4 ounces (small)

Place the large and small cups in front of the group.

What words can we use to describe these cups? (Blue. *Or*, Red. *Or*, Short. *Or*, Tall.)

Pour water into each cup.

What words can we use to describe how much water these cups could hold? (A little. *Or*, A lot.)

If you were really thirsty, what cup of water would you want? Why?

What if you just wanted a little sip of water, which cup could you choose? Which has less water?

What would you tell someone about how much water these cups hold?

We can use the words more *and* less *to describe how much something will hold.*

Will the small cup hold more or less water than the big cup? Why?

Make connections to using the words more and less when comparing other quantities as in Module 5.

Put the medium-sized cup on the table.

Turn and tell your partner: Will this hold more water than the small cup?

Have a student pour water from the small cup into the medium cup.

Do I have room for more water in this cup?

What does that mean?

Have a student pour more water in the medium cup.

Does this cup hold more water than the small cup?

Empty the large cup.

Turn and tell your partner: Will this large cup hold more water than the medium cup?

Have a student pour water from the medium cup into the large cup.

Do I have room for more water in this cup?

Have a student pour more water in the large cup.

Does this big cup hold more water than the medium cup?

Turn and tell your partner which cup holds less water than this big cup. (The medium one and the small one both hold less water.)

Empty the medium cup.

Pour the water from the large cup into the medium cup. When the medium cup is filled, stop and show students that there is still more water in the big cup.

Tell your partner: Does the medium cup hold more or less water than the big cup? How do you know?

What are some other containers that we use that can hold more than this big cup? (A bucket. *Or,* A wading pool.)

What are some other containers that we use that can hold less than this big cup? (A tiny cup. *Or,* A yogurt cup.)

We can use the words more *and* less *to label how much something will hold.*

Write *more* and *less* on the Math Talk chart and show a picture of a big cup and a small cup to illustrate the concepts.

Which cup holds more water? This student pours water from a smaller cup to a larger cup. The larger cup will hold more than the smaller cup. Pouring water helps students compare the capacity of the containers.

Chores

Students compare the size of containers and think about which container to use for different jobs.

Materials: a bucket, a pitcher, a large cup, and a small cup for small groups of students; one saucepan to show students as a reference

Write a list on the board or chart paper of chores you do when you get home.

When I get home, I have lots of jobs to do!

I put water in my hamster's bowl, put water in a pan to make macaroni and cheese for dinner, and give my dog a bath.

Give groups of students the set of containers (bucket, pitcher, large cup, small cup).

I need water for each of my jobs. I need help deciding which container to use for each job. I am going to tell you about my jobs and then you can help me decide which container I should use.

I need to give my dog a bath. He likes to take a bath in the bathtub. Which container should I use to fill up the bathtub?

Do I need a container that holds a lot or a little water? Talk to your partner and decide which container would be best.

Have students hold up the container they think is best (probably the bucket).

Why should I use the bucket?

Couldn't I fill the tub with the little cup?

Why would it take a long time to fill it?

Which holds more water, the bathtub or the bucket?

Put a check next to "give the dog a bath" on the list.

I need to make some macaroni and cheese for dinner. Which container should I use to put water in my pan?

Show the students the pan.

Should I use the bucket? Why?

Talk to your partner: What should I use?

> *It is true that any of the containers would work because you would just pour a little out of the container, but at this level students are likely not to think of it in that way. We can also talk about choosing the container that is closest to the amount of water we need.*

Have students hold up the container they think is best (probably a cup).

Which cup? Which cup holds more water?

Should I use the one that holds more water? Why?

Put a check next to "make dinner" on the list.

I have a hamster and I fill its little dish with some water each day. My son says it only needs a little bit of water.

Which container should I use for just a little water?

Have students hold up the container they think is best (probably the little cup).

Why?

Does this cup hold more or less water than the other cup?

Put a check next to "give the hamster water" on the list.

Review more and less by comparing the cups and buckets.

Show a container that holds less than this cup.

Show a container that holds more than this cup.

Show a container that holds less than this pitcher.

Goldilocks and the Three Bears

After listening to *Goldilocks and the Three Bears*, students decide which cup the bears would use.

Prepare a collection of small, medium, and large cups for groups of students to share.

Before Reading:

We are going to read a story about a baby bear, a mama bear, and a papa bear.

Listen for the size of the items that each bear has in the house.

During Reading:

Read *Goldilocks and the Three Bears* by James Marshall.

As you read, use other measurement vocabulary to describe items in the story (e.g., Baby Bear's bowl would hold less porridge than Papa's bowl. Papa Bear's chair is heavier than Mama Bear's.)

After Reading:

Give groups of students a collection of various-sized cups.

Find a cup that Baby Bear would use.

Have groups talk about it and then hold up the cup.

Why would Baby Bear use that cup?

Find a cup that Mama Bear would use.

Have groups talk about it and then hold up the cup.

Why would Mama Bear use that cup?

Find a cup that Papa Bear would use.

Have groups talk about it and then hold up the cup.

Why would Papa Bear use that cup?

In your math journals, draw a picture of the cup for each bear.

Label the cups as Baby, Mama, *and* Papa.

Use the observational notes page.

FORMATIVE ASSESSMENT

Use two different-sized cups.

Ask the student to show the cup that holds more water and explain.

Ask the student to show the cup that holds less water and explain.

Students should recognize and be able to explain which cup holds more or less water.

Record observational notes of student responses.

Additional Ideas for Support and Practice

The following ideas extend students' understanding of capacity and provide meaningful practice.

Filling Cups and Containers: Which Cup Holds More?

Students need lots of experience with filling cups and containers of all sizes with water, sand, beans, rice, or other simple contents.

Set up a water table or a large plastic storage container that students can use to explore and pour from one cup to another.

Place cards at the center asking questions such as:

Which cup holds more?
Which cup holds less?

To challenge some students, try prompts like:

How many small cups of beans does it take to fill the large cup with beans?

Cups with Partners

Partners explore the capacity of cups using *more* and *less*.

For each pair of students, have a set of 8–10 cups and containers of various shapes and sizes.

Player A selects two cups from the set and gives them to Player B.

Player A asks, "Which cup holds more?" and "Which cup holds less?"

Player B compares the cups and answers, then returns the cups to the collection and picks two different cups.

Players continue to pick cups and answer the questions, switching roles.

Note: You might have a container of beans available for students to check to be sure they are correct.

My Favorite Cup

Students bring a favorite cup from home and compare how much each will hold.

Have each student bring a favorite plastic cup from home.

Each student shares his cup and tells about it.

Then, students compare to see whose cup holds more or less.

Which cup holds less? Molly's elephant cup from the circus or Chuck's football team cup?

Students predict whose cup holds less, then fill that cup with beans.

They dump the beans into the other cup to see if the other cup can hold more beans.

Talk About It/Write About It

You are going to a movie and get to order a drink and popcorn.

Draw a picture to show the cups for your drink and popcorn. Which cup holds more? Which one holds less?

Vocabulary

heavy	*tall*
heavier	*taller*
light	*more*
lighter	*less*
long	*length*
longer	*height*
short	*weight*
shorter	

ONLINE RESOURCES

General Resources

Additional Problems

Measurement Student Observation Notes

Resources for Specific Activities

Balance Recording Sheet

Longer or Shorter Recording Sheet

These resources are available at http://hein.pub/MathinPractice, keycode MIPGK.

MODULE 11

Sorting and Classifying Objects

About the Math K.MD.3

Kindergarten students sort and classify objects, developing their skills at identifying ways objects are alike and different. Looking at objects in discriminating ways and talking about their attributes, or characteristics, builds students' skills at observing, comparing, and classifying. In kindergarten it makes sense to keep these sorting activities limited to ten or fewer objects in each group so as not to overwhelm students with too many objects to sort.

The key ideas focused on in this module include:

- **sorting items into groups**
- **describing how items are sorted**
- **adding an item to a sorted group.**

Talking about how they sort objects challenges students to identify characteristics that are alike and different and to develop the vocabulary to classify their groups and explain their sorting based on those characteristics.

- Are the objects the same color, size, or shape?
- What would they call the group: red objects, big objects, square objects?
- Are they comparing groups of objects based on how many are in each group?
- Are they comparing shapes based on the number of sides or the types of shapes (e.g., triangles or squares)?

Once students have identified similar characteristics and sorted the objects, they find the words to explain their sorts to their partners or the class. They decide on a title or label for each group that tells what they were thinking as they sorted the items.

Students refine their skills as they move from simple sorting tasks to more complex ones. Students sort items based on a given rule (sort by color) or they might sort items by attributes they have noticed themselves. Students might be challenged to add an item to a previous sort or justify why it belongs in a particular group. Throughout sorting activities, students talk about how they sorted and offer justifications about why the sort makes sense, both of which help refine their skills.

Lots of math connections can be made during sorting experiences.

- Students might count the number of objects in their groups, practicing their one-to-one correspondence as they touch and count each item and developing their understanding of cardinality as they name the quantity in the group.
- They might sort geometric shapes, explaining how they are alike and different, to practice their use of geometry vocabulary.
- They might sort objects based on length, talking about short and tall or heavy and light items.
- They might compare the quantity of items in sets to show their understanding of the progression of the counting numbers or comparing numbers.

Integrating sorting and classifying experiences with math content taught throughout the school year is an effective way to teach this skill. Sorting during geometry lessons, sorting while working on number skills, or sorting when exploring measurable attributes places sorting and classifying in a specific math context and allows our students to revisit this skill throughout the year.

Learning Goals

I can sort items into groups.

I can describe how I sorted.

I can add an item to a sorted group.

Exploring the Progression

PREVIOUS		NOW		NEXT
Preschool Possible experience with sorting by color and counting the items in a group	»	**Grade K** Sorting objects into groups; describing how they are sorted; adding an item to a group	»	**Grade 1** Using sorting to organize data; representing the number of objects or votes in a category by using tally marks

Lessons in This Module

Sort by a Rule
Sort by Color
Sort by Size

Sort and Describe
Sorting Attribute Blocks
How Did You Sort?
Add to the Sort

Ideas for Instruction and Assessment

SORT BY A RULE

Students sort objects by how they are alike or different based on a rule (e.g., sort by color or size) and then describe their sorts. They count the number of items in each set.

Sort by Color SHOW IT!

Students sort teddy bears by a given rule: color. Students will sort, count, and compare the groups of bears.

Materials: a bag of at least three different colors of teddy bear counters (or other similar manipulatives) with two or more of each color in the set (e.g., a set of four blue bears, three red bears, two yellow bears), pencil, and sticky notes for each student

Give each student a bag of bears.

Put the bears that are the same color together to make groups.

Observe as students work. They should be able to create groups based on color.

Let's see how you grouped your bears.

Who has a group of red bears? Blue bears? Yellow bears? Green bears?

When you grouped your bears by color, you sorted by color.

Write *sort* on the Math Talk chart.

Sorting means making groups of bears that are the same or groups of bears that are alike in some way.

How could you find out how many bears are in each group?

Count each group of bears and write the number on the sticky note. Then place the number next to the group.

Observe as students work.

Have students share something about their groups.

Tell me something about your bears by using a number and a color. You might tell me you have 3 red bears.

If you have yellow bears, how many do you have?

Let students share about the number of different colors of bears.

Can you look at your sets and tell your partner one color that has more bears than another color? Do you have more green bears than yellow bears? Tell your partner about your bears.

Listen as students compare the numbers of bears.

After they have shared with a partner, have a few share with the group.

How do I know you have more red than blue? How could you prove it to me?

Have them return the bears to the bags and switch bags with a partner.

Repeat, sorting by color, counting, and labeling with the number in each set.

After students have had several chances to sort and describe their bears, pose the questions:

I wonder which of you has the most red bears?

Who has more red bears than anyone else?

How could we tell?

Differentiation

Students who may need additional support may benefit from fewer items to sort into fewer categories. Starting with just red and blue teddy bears and telling students to sort by color helps them learn to differentiate in order to sort.

Differentiation

Some students may need assistance placing the same-color bears in small groups. Pieces of paper, coffee filters, cupcake liners, or paper plates can be helpful to provide space for students to create groups.

Differentiation

When counting the number of items in each group, some students might benefit from using a ten frame to organize the count.

Have students share how many red bears they have.

How will we know who has the most, more than anyone else?

Paolo had 6. Did anyone have more than 6?

Paolo had the most red bears.

I wonder which of you has the least blue bears?

Who has fewer blue bears than anyone else?

How could we tell?

Have students share how many blue bears they have.

How will we know who has the least, fewer than anyone else?

Erica had 1 blue bear. Did anyone have less than 1?

Erica had the least blue bears.

Write *most* and *least* on the Math Talk chart.

Have students turn to a partner and tell what it means to have the most or least of something.

This activity can be repeated several times with different manipulatives. Different types of manipulatives (e.g., bears, cubes, counters) can be combined to be sorted by color, adding the challenge that they are the same color but look different in other ways.

Sort by Size

Students sort teddy bears by a given rule: size. Students will sort, count, and order the groups of manipulatives.

Materials: a bag of manipulatives that can be sorted by size (e.g., small, medium, and large bears), pencil, and sticky notes for each student

What does it mean to sort?

How did we sort last time?

Give students bags of manipulatives.

Take the items out of your bag and look at them carefully.

What do you notice about the bears? (They are different colors. *Or,* They are different sizes.)

Another way to sort items is to sort by size.

How many sizes of bears are there?

What could we call them?

Put the bears that are the same size together. Don't worry about the color. We are not sorting by color today.

When you are finished, record the number of items in each set.

Allow students time to sort and label the sets.

How many large bears do you have? How many medium bears? Be ready to tell me something about your bears using a number and the size.

Allow students to share descriptions of their bears using numbers and size.

Look at your sets of bears.

Which one of your sets has the most bears in it? Turn and tell your partner.

Do you have the most small, medium, or large bears?

Be sure you tell your partner how you know it is the most.

Differentiation

Some students may work best with only small and large items and pieces of paper labeled "small" and "large."

Students are being asked to talk with precision.

Allow students to share their ideas with partners and then with the class, asking them to justify their choices.

Did any of your sets have the same amount?

How could you prove they are the same?

When finished, students switch bags of manipulatives and sort other sets.

Students may need to see and feel the shapes to sort and then color.

Use the observational notes page.

FORMATIVE ASSESSMENT

Give students a set of manipulatives that are different types and colors (e.g., a set of toy transportation vehicles, insects, or teddy bears in different sizes and colors).

Tell the student to sort by color.

Ask him to tell you the name of each group (e.g., red, blue, green).

Record observational notes of student work and responses.

If the student is not able to sort by color from those directions, ask him to make a set of things that are red and a set of things that are blue to see if he can sort when your directions are specific to the name of each set. Continue to offer sorting opportunities to the student, beginning with two groups before moving on to sorting into three groups. Have the student work with a partner during sorting tasks to allow him to see and hear the way others sort the objects.

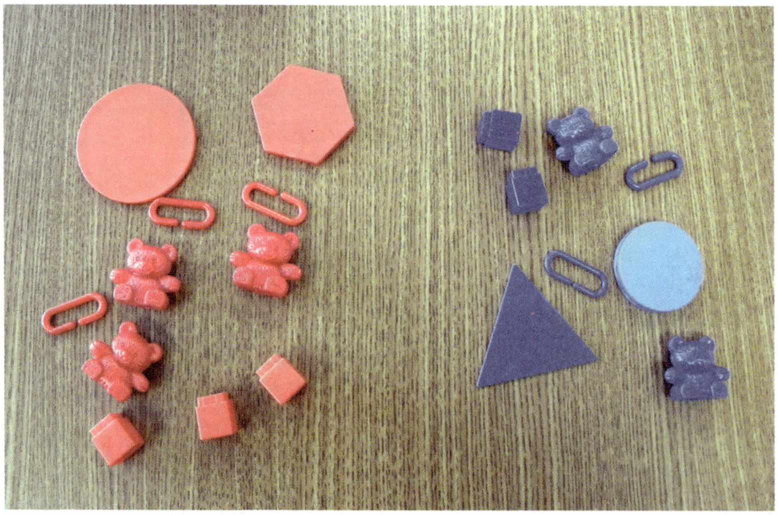

A student groups assorted materials by colors.

Additional Ideas for Support and Practice

The following ideas extend students' understanding of sorting objects by a given rule and provide meaningful practice.

Sorting Students

To model sorting, sort small groups of students at the beginning or end of lessons. Students can be sorted based on characteristics that include the following:

- shirt color
- eye or hair color
- tie, strap, or slip-on shoes
- tall or short socks
- bus number
- buying or packing lunch
- buttons or no buttons on shirts
- number of letters in the student's first name.

Begin by telling the students the rule that you used to sort the students.

Then, have students decide where classmates belong.

As they gain experience with sorting, ask the students to guess your rule.

Scoop and Sort

Students scoop manipulatives from a bowl and sort them.

Put a variety of manipulatives that vary based on size, color, and shape in a large bowl.

Students can use a ladle or large spoon to scoop out a set of items.

They sort the items and use a sticky note to label the categories and number in each category.

Sorting Rotations

In small groups, students sort items by a given rule.

Create four or five small-group rotations by placing a container of objects and an index card that lists a sorting rule at various tables around the room.

Divide students into small groups and send them to the tables.

Each group will sort the objects by the sorting rule.

Once groups have finished and their work has been checked by a teacher, they will clean up the area and move to the next table.

Repeat until all groups have visited every rotation area.

If students are not able to read the cards, set up examples or include drawings on the sort cards.

Some example rotations include the following:

- assorted socks: sort by size or pattern
- mixed vehicles: sort by land/water/air transportation
- buttons: sort by the number of holes
- farm animals: sort by type of animal
- attribute blocks: sort by color and shape.

Sort by Size

Student sort attribute blocks by size and then color shapes that are big and small.

Give students attribute blocks and tell them to sort by size.

Then give them the Sort the Shapes sheet to color big and small shapes.

Sort the Shapes

Color the **big shapes red.**
Color the **small shapes blue.**

Students apply their understanding of sorting by size to identify large and small shapes by coloring. For some students, it could be helpful to place the matching attribute block on the paper. Students may need to see and feel the shapes to sort and then color.

The Button Box

Students learn how a little boy sorted buttons and then sort their own buttons.

Before Reading:

Ask students where they might see buttons at home.

Discuss how some families keep a button box of extra buttons in case they lose a button.

During Reading:

Read *The Button Box* by Margarette S. Reid.

Point out different ways the boy is sorting the buttons (size, number of holes, material, etc.).

After Reading:

Give students a tray or mat and a handful of buttons.

Students work with a partner to sort the buttons by a given rule (e.g., number of holes, color).

After students have sorted their buttons by the rule, see if they can sort them a different way.

SORT AND DESCRIBE

Students extend their understanding of classification and begin to describe sorts. They sort any way they'd like and share their rule with others.

Thinking Through a Lesson

Sorting Attribute Blocks (SHOW IT!)

Students sort attribute blocks by color, shape, size, or thickness.

Materials: a bag of attribute blocks for each student

Give each student in the small group a bag of attribute blocks.

Put your blocks out in front of you.

What do you notice about them? (Different colors, shapes, sizes, and thicknesses.)

Take a few minutes to allow students to explore the blocks and talk about what they see. Depending on the sequence of your curriculum, students may not know all of the shape names, although they should be able to put like items together even if they struggle with the names of each shape.

Model the language for students as you talk specifically about the attribute blocks: large, small, thick, thin, shape names, and color names.

Students will most likely notice the differences in shape and size, but may not notice the thickness. If they do not, point out this attribute by holding up two blocks that have the same shape and size, but different thickness and ask students how they differ. Record the words thick and thin on the Math Talk chart.

SMP6

Show a way you could sort these blocks.

Allow students time to work, then have them turn and talk with a partner about how they sorted the blocks.

> If you use the term sorting rule, be sure to consistently remind students what you mean by that. What were they thinking about when they sorted the blocks?

> **Differentiation**
> Although we prefer to have students decide on a way to sort the blocks, some students may need suggestions to get started.

Tell your partner what sorting rule you used.

Tell your partner what you named each of your groups (Thick group and thin group. Or, Big and little group. Or, Blue and red group.)

Tell your partner how many items are in each of your groups.

Have a student share her sort. Ask her to tell the names of her groups (or how they are alike).

Ask questions like:

How many shapes are there in each of your groups?

Which group has the most? Or the least?

Encourage students to explain their thinking.

> Select a student to share who sorted by one attribute (e.g., color) and then select another student to share who sorted using a different attribute.

I notice that you have a big, thick, blue triangle with a small, thin, yellow triangle.

Why do you have them grouped together? (They are both triangles.)

What did you call your group? (Triangles.)

> Pose questions that encourage students to look at others' work and make sense of their thinking.

Have students answer questions about another students' sort.

SMP6

Why did Jenny put the big thin square with the big thick circle? (They are both big.)

Ask students to find another way to sort the blocks.

Have them share their new sorts with partners and have a few share with the group.

Continue to ask them for the names of the groups, how the shapes in the group are alike, and to tell the number of items in each of their groups.

> **Misconception**
> Some students may have difficulty sorting objects by multiple traits and may believe that there is only one way to sort their objects. Help students develop an understanding that objects can be sorted by more than one attribute by prompting them to find other ways to sort the objects

When interviewed, this student explains that he sorted his attribute blocks by color. Next he is asked to sort his blocks another way using a different sorting rule.

How Did You Sort?

Students sort various manipulatives and describe the rule used to sort them.

Materials: manipulatives that can be sorted in many different ways, such as shape, type, color, or size; a copy of the Sort Report recording sheet for each student

We have been sorting based on rules that you have been given.
Your job today is to create your own sort with your own rule for sorting.
We have manipulatives that can be sorted in many different ways.
Think about how you'd like to sort them.

Allow students to select the manipulatives to sort and give them time to work.

Some students may need help to select items or guiding questions to help them get started.

How do you plan to sort?
What items would you like to sort?
Would you like paper plates to sort your sets on?

Students could also self-select items to sort from the classroom collection.

Differentiation

As an extension, students can sort by two rules. Attribute blocks provide many opportunities for sorting with two rules. They might be sorted as large, blue shapes or shapes that are large and red. As students are working, observe how they select items and how they sort.

After students have sorted, they can record how they sorted, such as by size, color, or shape, and the number of items in each set, on the Sort Report recording sheet.

Have students share their sorts with the group.

What rule did you use to sort?

Which set had the least?

Did anyone else sort their items in the same way?

Continue to have other students share their sorts.

Have students walk around the room to observe the sorts.

Add to the Sort

Students use a variety of manipulatives to create a sort and add to a partner's sort.

Materials: a variety of items for students to sort, attribute blocks for you to demonstrate with

For more ideas on getting students to talk about math, see Math in Practice: A Guide for Teachers, Chapter 4.

Show the group some attribute blocks that are sorted as thick shapes and thin shapes. Have extra blocks off to the side for students to use.

Turn and tell your partner what rule I used to sort these attribute blocks?

How do you know?

It's your turn to add another block to the sort and explain why you added it.

Allow each student to select a block to add to one of the sets.

Which block did you add to the sort?

Why did you add that one to that group?

Next, have students work in pairs.

Have one partner use the manipulatives to create her own sort. Have her leave a couple items out of the sort so her partner can try to figure out where they go.

Partners work to figure out each other's sorting rule and try to add a shape to one of their partner's sets.

Talk with pairs as you watch them work.

How did Anna sort her items?

What did you add to her set?

Why did you add that?

Anna, did that work with your sort?

After one partner figures out the sort, have the other partner take a turn at it.

FORMATIVE ASSESSMENT

Give students the Add to a Sort assessment.

Have students draw a shape to add to each sort.

Students could be challenged to write the name of each group under the boxes.

Add to a Sort

Draw a shape in each box to add to the sorts below.

This student correctly adds a shape to each sort. Although it can be difficult for kindergarteners to draw shapes correctly, it is the concept of sorting that is being assessed, not the drawing of the shapes. If it is unclear what shape students are drawing, have them tell you what shape they are adding and why.

Additional Ideas for Support and Practice

The following ideas extend students' understanding of describing sorts and provide meaningful practice.

How Many Snails?

Students will identify groups of snails based on characteristics and count the number in each group.

Before Reading:

Students turn and talk to a partner about what it means to sort.

Tell students that you are going to read a book about snails that can be sorted.

During Reading:

Read *How Many Snails? A Counting Book* by Paul Giganti, Jr.

As you show each picture, ask students how they can sort the snails and how many snails are in each group.

After Reading:

Students draw pictures of snails that are sorted.

Have them label the pictures with the names of the groups and the number of snails in each group.

Sort Report

Students sort items, draw pictures to represent the sort, and label the categories.

Materials: collections of small toy cars, assorted plastic beads, or jewelry for students to sort

Students decide how to sort items:

- Vehicles can be sorted based on the number of wheels, color, size, or type.
- Jewelry or beads can be sorted by color, shape, or type.

Students use the Sort Report sheet to draw a picture of and label the categories and record the number of items in each group.

Guess My Rule

Students work with a partner to name each other's rule for sorting.

Materials: manipulatives that can be sorted in many different ways such as shape, type, color, or size

Player A creates a sort and asks Player B to "Guess my sort."

Player B names the sort.

Then players switch roles and continue with other sorts.

Other possible sorts to extend student thinking include:

- sorting by texture
- sorting by weight
- sorting by round/straight sides
- sorting objects by concept (for example clothes, school supplies, and animals).

Eliminate It

Use attribute blocks to show this set of shapes to a small group of students:

- large, thick, blue triangle
- large, thin, blue triangle
- large, thick, blue square
- small, thick, red triangle.

Which shape doesn't belong in the group and why?

Then, show this set to students:

- large, thick, red rectangle
- large, thin, red rectangle
- small, thin, red square
- small, thin, blue square.

Which shape doesn't belong in the group and why?

Spy the Sort Bags

Students identify items that belong and do not belong in sorting bags.

Materials: a gallon-size bag that contains two groups of manipulatives—one group that matches a sorting rule (for example, small yellow bears and small yellow fruit) and one group that does not match (for example, a small green bear, a large yellow banana, and a small yellow button)

The sorting rule for the bag must be established first (yellow/small).

Pass the bag around so that each student can look for what does not belong.

Ask students to tell what did not belong in the bag and why.

Then each object that does not belong must be explained using the sorting rules (e.g., the large banana does not belong because it is not the same size, but it does belong because it is the same color).

This activity can be turned into a math center.

Provide students with empty bags and a bucket of various manipulatives.

Students can make up bags of items and switch with a partner to find what does not belong.

Vocabulary

categories	rule
groups	set
least	sort
most	

ONLINE RESOURCES

General Resource

Additional Problems

Sorting Student Observation Notes

Resources for Specific Activities

Add to a Sort Recording Sheet

Sort Report Recording Sheet

Sort the Shapes Recording Sheet

These resources are available at http://hein.pub/MathinPractice, keycode MIPGK.

MODULE 12

Introducing Geometry

About the Math K.G.1; K.G.2; K.G.3

Many kindergarten students appear to already know basic shapes, like squares, circles, triangles, and rectangles, but with a little probing we find that they identify shapes simply by their appearance (it looks like a triangle). So although they may call this a triangle △, they do not call this ▽ a triangle .

In kindergarten, students learn to identify and name shapes regardless of their size or orientation. A triangle is a triangle regardless of how it is positioned (see Figure 12.1).

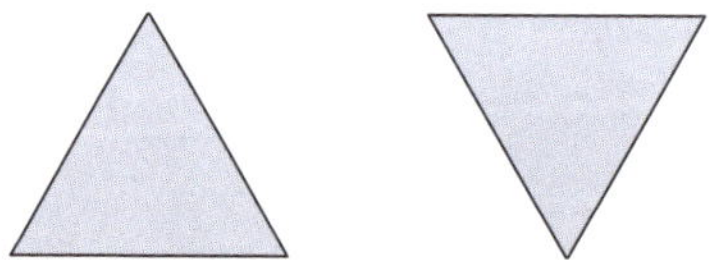

Figure 12.1 Despite their orientation on the page, both of these shapes are triangles.

The key ideas focused on in this module include:

- distinguishing between flat and solid shapes
- naming and describing flat and solid shapes
- accurately using words that show position like *above*, *below*, *next to*, *in front of*, and *behind*.

Truly understanding shapes and being able to describe shapes means that children must know the attributes of those shapes. What makes a triangle a triangle? What makes a rectangle a rectangle? Students sort and categorize shapes as they observe their attributes (see Figure 12.2). They notice and talk about such things as curved or straight, number of sides, or number of corners or vertices. They talk about similarities between the shapes, reasoning that some things matter (like the number of sides for a square), but others don't (like the color of a square).

Figure 12.2 Students sort shapes and talk about what they see, finding the ways that all squares or all triangles are alike.

Students learn that shapes can be two-dimensional (flat) or three-dimensional (solid). They touch and observe the shapes to better understand them. They notice cubes have flat sides, but spheres roll. They notice that solid shapes have surfaces that are flat shapes (e.g., the face of a cylinder is a circle). Experiences examining flat and solid shapes, and looking for examples of the shapes in the world around them, help students begin to make sense of and reason about shapes.

At this level, students do not need to know the terms two-dimensional and three-dimensional, but use the more common terms of flat and solid shapes. The focus at this level is on understanding the differences between these types of shapes, not on knowing the words two-dimensional and three-dimensional.

Working with positional language like *above*, *below*, *in front of*, *behind*, and *next to*, provides students with the words that are necessary to describe objects in their environment. Through hands-on experiences in which they place objects in various positions, as well as lots of math talk familiarizing them with these words, students become able to use the words as they describe objects.

Learning Goals

I can name and tell about flat and solid shapes.

I can tell which shapes are flat and which are solid.

I can use words that show position, like above, below, next to, in front of, *and* behind.

Exploring the Progression

PREVIOUS	NOW	NEXT
Preschool Possible knowledge of some basic two- and three-dimensional shapes	**Grade K** Identifying and naming shapes regardless of their orientation; distinguishing between and naming flat and solid shapes; using words that show position (e.g., *above, below*)	**Grade 1** Building or drawing shapes based on attributes; combining two- or three-dimensional shapes to form composite shapes; refining understanding of vocabulary terms (e.g., *faces, vertices, edges*)

Lessons in This Module

Ideas for Instruction and Assessment

UNDERSTANDING SHAPES BY ATTRIBUTES

Students learn to identify shapes based on their attributes through hands-on explorations, sorting activities, and discussions.

Introducing Shapes with Attribute Blocks

Students are introduced to flat shapes based on the number of sides and explore some foundational shape vocabulary.

Give each student a set of five thin attribute blocks (one each of circle, square, triangle, rectangle, and hexagon).

Have students lay their shapes on the table in front of them.

You each have five shapes.
I am going to ask you to find shapes for me.
You can help your partner if he or she needs help.
Show me the triangle.

Have students hold the triangle in the air.

How did you know that was a triangle?
What does a triangle look like?

Show students what you mean by a *side* and have them count the sides of their shape.

How many sides does your triangle have?
How many sides does your partner's triangle have?
Does everyone's triangle have three sides?

Record the word *triangle* on the Math Talk chart, draw a picture, and write *3 sides*.

Have them put the triangle back.

Show me the rectangle.

Have students hold the rectangle in the air.

How many sides does your rectangle have?

How many sides does your partner's rectangle have?

Does everyone's rectangle have four sides?

Record the word *rectangle* on the Math Talk chart, draw a picture, and write *4 sides.*

Have them put the rectangle back.

Show me the circle.

Have students hold the circle in the air.

How many sides does your circle have?

What does a side look like? Is it straight or curved? Look at the sides of your triangle and rectangle. Are they straight or curved?

Show the sides on the triangle and rectangle again.

Sides on a shape are straight. Does this circle have any sides?

Does everyone's circle have zero sides?

Show students what you mean by straight by drawing a straight line and a curved line on the board.

Record the word *circle* on the Math Talk chart, draw a picture, and write *0 sides.*

Have them put the circle back.

Show me the square.

How many sides does your square have?

Didn't we have another shape with four sides?

What do you notice about the sides of your square? Are some short and some long?

They don't look short and long, they look the same. The sides are the same length.

Are the sides the same length on your partner's square?

Record the word *square* on the Math Talk chart, draw a picture, and write *4 sides that are the same.*

Continue with the hexagon (six sides) talking about the sides and recording the information on the Math Talk chart.

When all of the shapes are on the chart, name random shapes (e.g., rectangle) and have students pick up that shape and tell their partner how many sides the shape has.

On another day, repeat this lesson but focus on corners or vertices in each shape.

See connections to sorting in Module 11.

Remove the thick blocks from the set so the blocks show two-dimensional shapes. Students do not need all of the thin blocks, just enough to sort is fine.

You may need to introduce students to the term attribute blocks as the name of these types of shapes so you can then refer to them as blocks. Students often think of blocks as cubes or rectangular prisms and using the term without mentioning that these look a bit different may be confusing to them.

Include comments about corners/vertices if that term has been introduced.

Sorting Attribute Blocks by Shape

Students sort attribute blocks to discuss attributes of flat shapes.

Give each pair of students a set of 10–12 assorted thin attribute blocks (circles, squares, triangles, rectangles, and hexagons).

What shapes do you see when you look at the blocks?

Work with your partner to sort your blocks by shape.

Observe as students sort the shapes.

Have students share how they sorted the blocks. Are all of the squares together or are they putting small squares in a different group than large squares?

After students have sorted blocks, ask:

How did you sort your blocks?

What did you name your groups? (Circles, triangles, squares, rectangles, or hexagons.)

Do the shapes in each group have the same number of sides?

Have students point to the group that has three sides.

What do we call those shapes?

Have students point to the group that has six sides.

What do we call those shapes?

Have students point to the group that has four sides that are the same length.

What do we call those shapes?

If I gave you another shape with three sides, where would it go? (With triangles.)

Then give each student an additional attribute block.

Which group does this block belong in? How do you know?

Have students add that shape to their sort and tell the class why they put it in that group.

Thinking Through a Lesson

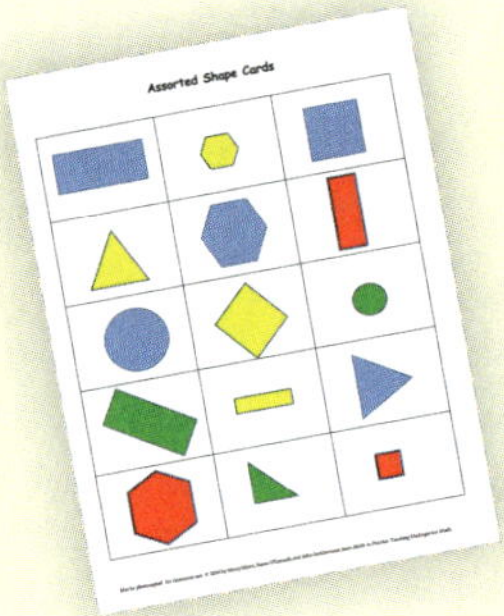

Shapes in Different Orientations

Students sort and name shapes, then discuss shapes that are in different orientations.

Give each group of three students a set of shape cards with circles, squares, triangles, rectangles, and hexagons in various sizes and orientations.

Have the students sit in groups of three.

Give each group a set of cards.

Look at the cards in front of you.

What do you see on them? (Shapes. *Or,* Circles, squares, triangles, rectangles, and hexagons.)

Work with your group to sort the cards by shape.

If students simply say shapes, ask them to be more precise by using the shape names.

SMP6

Observe as students sort the shapes.

When groups are finished sorting, say:

What did you notice about the shapes on the cards? (They are different shapes. *Or,* Some are the same shape but some look different. *Or,* Some are different colors. *Or,* Some are big and some are small. *Or,* Some are tilted.)

Did you have any shapes that you couldn't figure out where to put them? (Yes.)

If you did, it's okay, we'll find a place for them as we talk about the shapes together.

The focus of the discussion should be on the attributes for each shape. Discuss that attributes, such as size and color, do not matter when you name the shape.

Begin to discuss the shapes, starting with the group of squares.

Have each member of the group hold a card with a square.

How are these shapes the same? (They have four sides, four corners, and all sides are the same length.)

SMP6

Students holding the cards keeps them engaged and focused on each shape as you discuss it.

Can a square be big or little? (Yes, they are still squares.)

Does it matter what color the squares are? (No, they are still squares.)

Push students to be specific. Rather than "they match" have them tell you how they match. If they say they are squares, have them tell you how they know.

What about this one that looks like it is turned?

Show the card with the square that is turned with vertices at the top and bottom.

Is this a square? (Students may disagree.)

How do you know if a shape is a square? (It has four sides and four corners and the sides are the same length.)

Does this shape have four sides that are the same length and four corners? (Yes.)

Turn the card a bit. Does it look like a square now? (Yes.)

It was always a square! It doesn't matter how it is tilted, it is still a square.

Put all of your squares back together.

Although the orientation changes the look of a shape, the important understanding is that it does not change the shape itself. It is a square regardless of how it is positioned. Students need to be able to manipulate shapes to understand this, which is why they are exploring with shapes on cards that can be turned to different positions.

SMP3

Continue with the group of triangles.

What can we name this group? (Triangles.)

How are these shapes the same? (They have three sides and three corners or vertices.)

Does it matter what size the triangles are? (No, they are still triangles.)

Does it matter what color the triangles are? (No, they are still triangles.)

What about this one that looks like it is turned? (It is still a triangle with three sides and three corners.)

Turn the card a bit. What do you notice? (It looks like the other triangles.)

It is a triangle because it has three sides and three corners.

Put all of your triangles back together.

Ask how rectangles are the same as or different than squares.

Continue with circles, rectangles, and hexagons.

Students can copy the names of each shape or create small preprinted labels for students to put in their journals.

In their journal, students then draw examples of triangles, squares, circles, rectangles, and hexagons.

SMP4

Tell students to show two examples of each shape, reminding them that the shapes can be big or small, any color, and tilted in any way as long as they have the right number of sides and corners.

You can split this lesson into shorter lessons in which you explore different shapes on other days. Revisiting the ideas over several days can be helpful for many students.

Class Shape Chart

Work together to review the attributes of shapes by creating a Class Shape chart that includes defining attributes.

Prior to the lesson, begin the shape chart with three columns: Shape Name, What We Look For, Examples.

Shape Name	What We Look For	Examples

Distribute the shape cards used in the previous lesson (Shapes in Different Orientations).

Have students sort the cards.

Give students individual dry erase boards.

What's the name of one of the shapes that you sorted?

Squares, great! Let's start with squares.

Show the shape chart to students and tell them they are going to help you create a chart about the shapes they sorted.

Add *square* to the Shape Name part of the chart.

Tell your partner: What makes a square a square? What do you notice about all squares?

Have students share with the class. Add correct student responses to the chart. Use misconceptions to clarify student thinking.

Hold up an example of a square from your set of cards.

Have students draw a square on their dry erase board and show their partner.

Add several examples of squares on the chart in various sizes, colors, and orientations.

Tell your partner: What makes a triangle a triangle?

Have students share with the class. Add correct student responses to the chart. Use misconceptions to clarify student thinking.

Hold up an example of a triangle from your set of cards.

Have students draw a triangle on their dry erase board and show their partner.

Draw several examples of triangles on the chart in various sizes, colors, and orientations.

Continue with other shapes.

Students can draw shapes in their journals as the class chart is being made, which can increase engagement during this lesson.

Shape Name	What We Look For	Examples
Square	Four sides all the same length Four corners	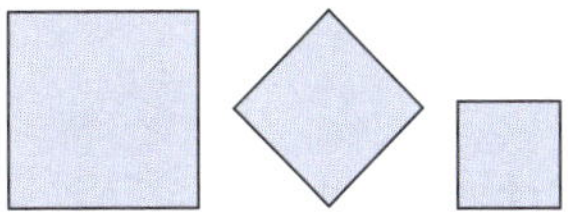
Triangle	Three sides Three corners	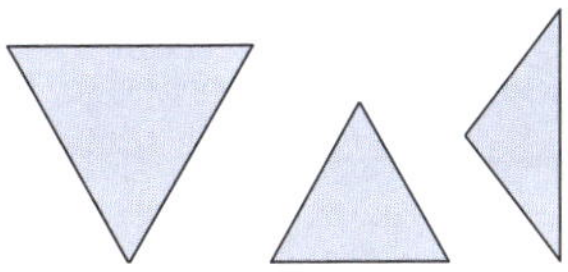

Shape Name	What We Look For	Examples
Rectangle	Four sides Four corners	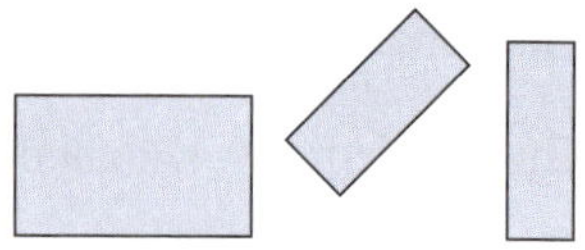
Circle	No sides or corners	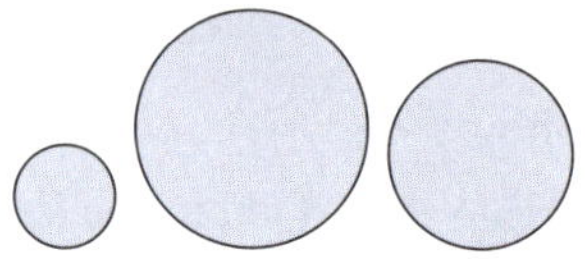
Hexagon	Six sides Six corners	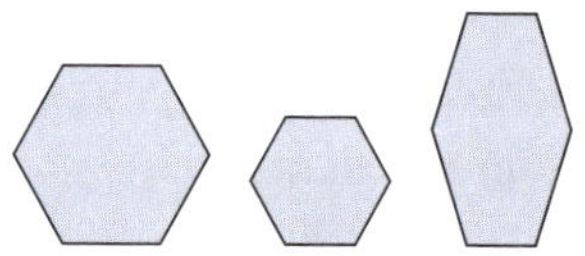

Students could be assessed using shape cards to include nontraditional orientations.

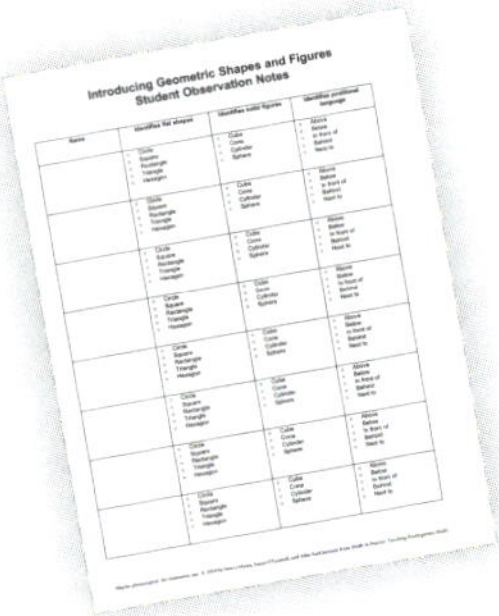

FORMATIVE ASSESSMENT

Use the thin attribute blocks to assess that students can name each shape.

Show the student a circle, square, rectangle, triangle, and hexagon and tell him to name each shape.

Record observational notes of student responses.

If the student is not able to recall the name of each shape, place the shapes in front of the child.

Say, "Touch the circle. Touch the square. Touch the rectangle. Touch the triangle. Touch the hexagon."

Use the observational notes page.

It is easier for some students to identify the shapes rather than recall the names.

Make a note of any shapes the student cannot recognize or name. Provide additional experiences with those shapes.

Additional Ideas for Support and Practice

The following ideas extend students' understanding of two-dimensional shapes and provide meaningful practice.

Round Is a Mooncake

Students begin by identifying shapes in a book and then go on a shape hunt around the school.

Before Reading:

Review the shapes on the Class Shape chart.

Tell the students that you will read a story about a little girl who finds shapes in her neighborhood.

During Reading:

Read *Round Is a Mooncake* by Roseanne Thong.

Discuss the shapes in the illustrations.

After Reading:

Fold pieces of paper into fourths and label each section as *circle*, *square*, *rectangle*, and *triangle*.

Refer to the Class Shape chart for examples of each shape.

Have students hunt for the shapes around the classroom, hallways, or in other areas of the school. When they find a shape, they draw a picture of what they found in that section of their paper.

Talk about the shapes that students found.

I notice that you drew a picture of the clock. What shape is that?
How could you tell it was a circle?
Did anyone else draw the clock as a circle?
What else did you find that is a circle?

Have students share other examples of circles.

What did you find that is the shape of a square?
What makes a square a square?

Have students share examples of squares.

What did you find that is the shape of a rectangle?
What makes a rectangle a rectangle?

Have students share examples of rectangles.

Triangles can be hard to find. Did anyone find a triangle?

Have students share examples of triangles.

Shape Sort

Students sort cards showing shapes in different orientations.

Cut out sets of shape cards with circles, squares, triangles, rectangles, and hexagons in various sizes, colors, and orientations.

Students sort shapes based on names and attributes.

Students then draw and label the shapes in their math journal.

Common Error

If a student has difficulty identifying a shape in a nontraditional orientation, turn the shape until it looks like a traditional shape (e.g., if a triangle is "upside down," turn it for the student until the triangle is on a flat side. Ask the student, "What shape do you see now?" Return the triangle to the original position and ask: "Is this the same shape? What shape is this?"

Talk About It/Write About It

Use attribute blocks to show students examples of squares and rectangles.

How are squares and rectangles alike?

How are they different?

Students can draw examples of both and label.

What is a triangle? How do you know a shape is a triangle?

What is a square? How do you know a shape is a square?

Geoboards

Students create shapes using geoboards.

Students use rubber bands on geoboards to make shapes.

At first give students free range to make any shape they wish.

Then have students make specific shapes that you name (e.g., triangle).

Finally, give attribute clues and have students make the desired shape (e.g., three sides).

This student works to create different triangles on a geoboard.

Attribute Clues

Students use thin attribute blocks to identify characteristics of the shapes.

Each student picks an attribute shape and sits on the carpet in a circle.

Give a clue (e.g., three corners, four straight sides, no corners).

Students stand up if the clue applies to their shape.

The students name the shape.

Then, another clue is given and those students stand.

Play continues until all students have heard an attribute that applies to their shape.

Have students select a different shape and play again.

UNDERSTANDING SOLID FIGURES BY ATTRIBUTES

Students examine solid figures, learn how they differ from flat shapes, and explore their attributes.

Introducing Solid Figures

Students examine sets of three-dimensional figures to identify their characteristics and names.

Each pair or group of four will need a set of flat shapes and a set of solid shapes (e.g., cube, sphere, cylinder, cone).

Put out some flat shapes (e.g., circles, squares, triangles, rectangles, hexagons).

Tell your partner the names of each shape and what you know about the sides and corners of each shape.

Listen as students share their ideas, then have them share with the class.

Put a set of solid shapes at each table with the flat shapes.

How are these shapes different from the ones we have been working with?

What do you mean by bigger or higher?

We call circles, triangles, squares, rectangles, and hexagon flat shapes because they are flat like paper.

Place a few of the shapes (flat and solid) near the edge of the table and have students squat down and look at the shapes, looking at them with their eyes at table level.

What do you see? (They stick up and we can see them but the flat ones don't.)

These shapes, that stick up, we call solid shapes.

Hold up a cube.

Find the shape that looks like this one. You might hear me call these shapes figures. It is another way to say shapes.

What do you notice about this figure? (It looks like a box.)

These flat parts (show a face) are called faces. *Not like our faces, but the same word.*

What do the faces of a cube look like? Do they look like any shape you know?

Giving students an eye-level perspective helps them see the height of the three-dimensional shapes.

The vocabulary in geometry is pretty intense and students need lots of exposure to these new words to internalize them. Use the terms, frequently reminding students of the meanings, and post the words and picture clues on word walls and Math Talk charts.

For more ideas on developing vocabulary, see Math in Practice: A Guide for Teachers, *Chapter 4.*

Have students examine the faces.

What shape are they?

What do you know about a square?

A square is a flat shape, but a cube is a solid figure because it sticks up. It is not flat.

Does your cube roll? Try it.

Does your cube stack? Try it.

Why does it stack?

If you are holding the cube for your group, bring it up to the front.

Display the assorted cubes from each group.

Look at these cubes.

What do you notice? (There are different sizes and colors. They all have sides that look like squares.)

Label the set as *cubes*.

Continue with cylinders.

In your set of solid figures, can you find one that looks like a can?

It could be a soup can, a soda can, a can of cat food, or a can of fruit.

Hold up a figure that looks like a can.

This is another solid figure called a cylinder.

Does this cylinder have a face? Remember faces are flat parts of the shape.

What do you notice about the faces of a cylinder?

Turn it in your hand. What do you notice?

What happens if you lay it on its side?

If you lay the cylinder on its side, it can roll like a rolling pin. You could use it to roll out clay.

What if you set it this way (set it on one of the faces). *Can you stack cylinders one on top of the other?*

Let's group the cylinders together.

Put the cylinders in a group in front of the students. Ask them to tell a partner what they notice about all of the cylinders.

We have cylinders that are tall and skinny, tall and fat, or short and fat, but they are all cylinders.

Label the set as *cylinders*.

If needed, do this lesson in two parts to slow down the introduction of the figures and related vocabulary. You might introduce two figures, then revisit those the following day and introduce one or two more.

Continue with cones.

Can you find a figure that would be called a cone?

Hold it up.

Why did you think that would be called a cone?

Does the cone have any faces?

What does the face look like?

Can you roll the cone?

Can you stack them one on top of the other?

Bring your cones up to make a group of cones.

Put the cones in a group in front of the students. Ask them to tell a partner what they notice about all of the cones.

They have different sizes and colors, but they are all cones.

Label the set as *cones*.

Continue with spheres.

You should have one solid figure left.

What do you have?

Yes, it looks like a ball. The name of the ball is a sphere.

All round balls are spheres.

What types of ball do you know that are spheres?

Does your sphere roll?

Can you stack spheres?

Let's make a set of spheres.

Look at the different sizes and colors of the spheres. They are all spheres.

Label the set as *spheres*.

This student explores solid shapes to see if they will roll or stack.

Real Shapes in the Classroom

Students explore real-world examples of solid figures in this small-group lesson.

Some real-world examples of solid figures include the following:

- cubes: tissue boxes, crates or bins, dice, Rubik's Cubes®
- spheres: balls, globes, marbles, tree ornaments
- cones: party hats, ice-cream cones, playground cones
- cylinders: canned goods, pillar candles, batteries, soda cans, drums

Place a set of assorted items at each table.

These are real objects but they are shaped like the solid figures we have been talking about.

Hold up a ball.

What figure is this?

Do you see any other spheres?

Let students sort through the items to find spheres.

How are they different?

How do you know they are spheres?

Hold up a can of soup.

What figure is this?

How do you know?

Do you see any other cylinders?

Let students sort through the items to find cylinders.

How are the cylinders different?

How are they alike?

Continue having students locate real-world objects that are shaped like cones and cubes.

Have them talk about how the real-world objects are alike and how they are different.

Begin a class list of real-world objects that are shaped like cubes, spheres, cylinders, and cones.

Have students draw a real-world object for each solid shape in their math journals.

FORMATIVE ASSESSMENT

Use a set of solid figures to assess whether students know the names of each.

Show the student a sphere, cube, cone, and cylinder, one at a time, and ask her to name each.

Record observational notes of student responses.

If the student is not able to recall the name of each figure, place the figures in front of the child.

Say, "Touch the sphere. Touch the cube. Touch the cone. Touch the cylinder."

It is easier for some students to identify the shapes rather than recall the names.

Use the observational notes page.

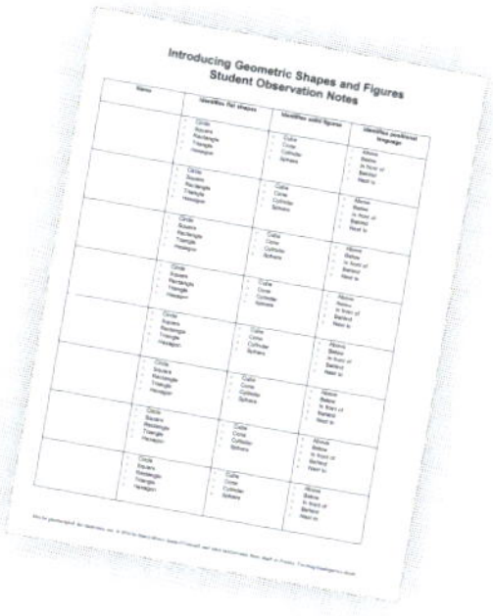

Faces of Solid Figures

Students examine the faces of solid figures.

This activity can focus on one figure at a time, over a few days, to avoid confusion between the figures and new vocabulary terms.

Give each student (or partners) a set of solid figures and have a set for teacher demonstration.

Pick up a cube.

What do we call this figure?

Show me your cube.

What do we call the flat part of a solid shape?

How do I know which parts are flat?

Rub your hand along the flat side. Allow students to touch the flat side.

Is it curved?

Would it lay flat on the table?

Show how the face of your cube lies flat on the table.

What shape do the faces of your cube look like?

I need a way to remember that these flat parts are called faces.

Put a smiley face sticker on each face of the cube.

Do some of the other shapes have flat faces, too?

Work with your partner. Look at your figures and see which ones have faces. When you find a face, put a smiley face sticker on it so we remember it is a face.

How will you know if it is a face?

Observe as students work with partners to find faces and put stickers on them. If students do not seem to understand, show them one face, have them feel the flat surface, and then ask them to find another.

Which figures have faces that look like circles?

Show me your cylinder. How many faces does it have? Count them with your partner.

Show them the two faces on your cylinder.

Show me your cone. How many faces does it have? Count them with your partner.

Show them the one face on your cone. What shape does it look like?

Show me your sphere. Does it have any faces?

Why not?

Show me your cube. Does it have any faces?

How many does it have? Count them with your partner.

If students have difficulty counting the faces, they can use their pencil to put a check or dot on the sticker as they count it.

Count them on your cube: one on the bottom and one on top (one, two . . .). Then count around the cube: three, four, five, six.

What shape do the faces look like? (Squares.)

Smiley face or dot stickers can be used to label the faces of figures.

Additional Ideas for Support and Practice

The following ideas extend students' understanding of solid figures and provide meaningful practice.

Solid Shape Hunt

Students hunt for solid figures and draw pictures.

Fold pieces of paper into fourths and label each section as *cube, cone, cylinder, sphere.*

Have students hunt for the figures around the classroom, along hallways, or in other areas of the school.

When they find the figure, they draw a picture of what they have found.

Discuss the figures that students found.

I notice that you drew a picture of a ball. What figure is that? How could you tell it was a sphere?

Did anyone else draw a ball as a sphere?

What else did you find that is a sphere?

What did you find that is the shape of a cube?

What makes a cube a cube?

What did you find that is shaped like a cylinder?

What do you know about a cylinder?

Cones can be hard to find. Did anyone find an example of a cone?

Mystery Figure in a Bag

Students feel shapes hidden in a bag and guess the mystery shape.

In a small group of students, put a figure in a brown paper bag and have each student feel it. After everyone has had a turn, ask questions such as these:

What figure is in the bag?

How do you know?

What did you feel?

Does anyone agree or disagree?

Repeat with other figures.

Trace the Face

Students trace the face of some solid figures.

Place several different-sized cubes, cylinders, and cones at the center.

Have students trace the face(s) of each.

What do you notice?

What flat shapes did you make?

Students label the faces as *square* or *circle*.

Students can turn their tracing into an art project, coloring the shapes of the faces they have traced.

Students can also use tempera paint and solid shapes. They dip the shape into the paint and stamp the shape of the face on their paper.

Stack, Slide, and Roll

Students explore whether solid figures can stack, slide, or roll.

Partners have a set of solid figures and a paper folded in thirds with the following labels on the sections: *Stacks, Slides, Rolls.*

Students work together to find the shapes that stack and then write them (or draw them) in that section of the paper.

Then, they find shapes that slide and write or draw them in that section of the paper.

This means to slide along the table without rolling (e.g., a flat face will slide on the table but a sphere rolls).

Finally, they test the shapes to see if they roll and then write or draw them in that section of the paper.

Shape Riddles

Create riddles to get students thinking about the attributes of each shape and figure. This helps students know the difference between flat shapes and solid figures.

Pose the riddles and have students turn and talk with a partner to solve each one.

I am thinking of a flat shape that has three sides and three corners. What is it?

I am thinking of a solid shape that has two faces shaped like circles. What is it?

I am thinking of a flat shape that has four sides that are all the same length. What is it?

I am thinking of a solid shape that has no faces. What is it?

What Doesn't Belong?

In this small-group activity, students examine bags of shapes to find the one that doesn't belong and justify their decision. This activity includes flat and solid shapes.

Prepare bags of shapes (in assorted colors and sizes) in which one shape does not belong with the others. Examples:

- four triangles and one square
- three hexagons and one circle
- five cubes and one square
- three squares and one triangle
- four cylinders and one cone
- three rectangles and one square

Have students dump out the bag and decide what does not belong.

Students then tell why it does not belong.

Ask students to use math words as they explain their thinking (e.g., one is a triangle and the others are hexagons; one has four sides and the others have three sides; one is a flat shape and the others are solid shapes).

Which One?

Students show their understanding of attributes as they answer questions about flat and solid shapes.

In small group, give each student two shapes (two- or three-dimensional).

Ask questions and when students have a shape or figure that answers the question, they hold it up.

Then, the group discusses the shapes or figures that are displayed. Sample questions:

Which shape has four sides?

Which figure has faces that are squares?

Which shape has three corners (vertices)?

Which shape is flat?

Which shape has sides of equal length?

Which figure has faces that are circles?

Which figure is solid?

UNDERSTAND POSITIONAL WORDS

Students explore positional words like *above*, *below*, *in front of*, *behind*, and *next to*, through movement, explorations, and discussions.

Develop Positional Terms

Students learn positional language by identifying items and their position in the classroom.

Prepare vocabulary cards with these words: *above*, *below*, *in front of*, *next to*, and *behind*.

What is above the board (e.g., the clock or alphabet)?

What does it mean to be above?

The vocabulary cards could include an illustration for the word.

Show the students the vocabulary card with the word *above*.

Have students point above their heads to demonstrate *above*.

When we hear or say the word above, *we will point above our heads to show above.*

Who can use the word above *to describe where something is in our classroom?*

Have students share some ideas.

As students give examples, move around the room to point out the items that are above.

Movement keeps the students actively involved in the lesson and reinforces new vocabulary words as they are introduced.

Continue with *below*:

What is below the board (e.g., posters or charts)?

What does it mean to be below?

Show the vocabulary card with the word *below*.

Have students point down to the ground to demonstrate *below*.

Who can use the word below *to describe where something is in our classroom?*

As students give examples, move around the room to point out the items that are below.

Differentiation
Some students may benefit from being shown the vocabulary cards or using the pointing signals from the introductory lesson.

Continue with other positional language.

Show the vocabulary card with each new word.

Have students point in front of, behind, and next to themselves as new vocabulary words are introduced. Remind students that *next to* can be to the right or left of an item.

Positional Puppets

Students use classroom puppets or stuffed animals brought from home to demonstrate positional language.

Have students sit in chairs to use puppets or stuffed animals to demonstrate positional language.

Put the puppet above your head.
Put the puppet below the chair.

Observe as students position the puppet.

Put the puppet in front of you.
Put the puppet behind the chair.
Put the puppet next to the chair.

Place the puppet above your head.

This puppet is below my head.
Do you agree or disagree? Why?
Where is the puppet?

Parking Cars

Students practice positional language by placing toy cars in specific locations.

Give each student a small toy car and a paper bowl.

Have students turn the bowl upside down on the table.

Have them listen to directions for parking their car in different positions.

Park your car above the bowl.

Use synonyms whenever possible. Talking about being above, or on top of, the bowl helps students see the connections between these terms.

Watch as students place the cars on top of or above the bowl.

Stop and talk about any directions that students misinterpret.

Park your car below the bowl.

Look for students to lift up the bowl and place the car below it.

Park your car behind the bowl.
What does it mean to be behind something? When I look at the bowl, will I be able to see the car very well? Why not?

For some terms like behind and in front of, it is important to establish where the front and back are. Taking time to talk about that at the start of the task minimizes confusion with these terms.

Ask students to park their car next to the bowl.

Why did some of you park the car on one side and some of you park the car on the other side?

Ask students to park their car in front of the bowl.

What does it mean to be in front of something? What would I see if I look at the bowl and your car if your car is in front?

Have students work with partners, giving each other directions for parking the cars.

This student parks her car in front of the bowl.

Positional Language with Solid Figures

Students listen to directions to position solid shapes.

Give each student a set of solid shapes.

Have them listen to directions to place the shapes in specific positions.

Show me the cube and the cylinder.

Put the cube above the cylinder.

Observe as students move figures to the correct location.

Which shape is above?

Have students put both shapes back.

Show me the cube and the sphere.

Put the cube in front of the sphere.

Common Error

Students may make errors in the positioning because they misunderstand the positional language or the name of the figure. If students are having difficulty with these tasks, have them tell you the shape names and then give the directions to try to isolate why the error is occurring.

Have students turn and tell a partner which shape is in front.

Have students put both shapes back.

Show me the cube and the cylinder.
Put the cylinder below the cube.

Have students turn and tell a partner which shape is below.

Continue with similar directions.

FORMATIVE ASSESSMENT

Option 1

Read the directions to students:

Draw a dot on the shape next to the star.
Color the shape below the star.
Put an X on the shape above the star.

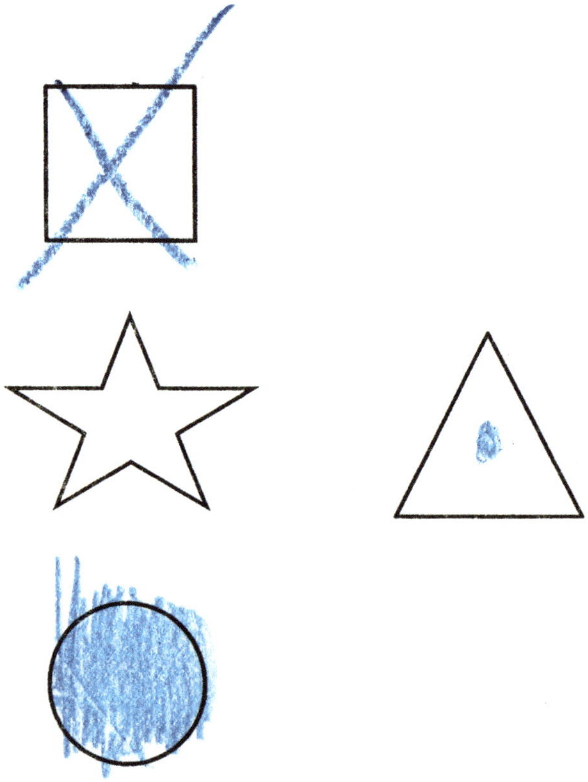

This student follows directions to demonstrate his understanding of positional language.

If students do not show an understanding on this written task, interview them, asking them to point to the shape next to the star, below the star, and above the star.

If they are still have difficulty with the paper-and-pencil task, give students an item and have them put it next to their seat, below their chair, and above their head.

Make notes on specific directional terms that may be problematic.

Option 2

Conduct an interview assessment using a cylinder, a cube, and a sphere.

First, have the student identify the figures.

If he is unable to identify the figures, do not continue with the assessment. Provide additional exposure to the solid shapes.

If he is able to identify the figures, observe if he is able to correctly position the shapes for these prompts:

Put the cylinder above the cube.
Put the cube in front of the sphere.
Put the cylinder below the cube.
Put the cylinder behind the sphere.
Put the cube next to the sphere.

Be sure to clarify front and back so you can correctly interpret if students know *in front of* and *behind*.

Use the observational notes page.

Record observational notes of student responses.

Additional Ideas for Support and Practice

The following ideas extend students' understanding of positional language and provide meaningful practice.

Move It

Students move objects using the interactive whiteboard to show their understanding of positional words.

Display a scene on the whiteboard (e.g., house, tree, and swing set).

Place some items for students to drag to different positions in the scene (e.g., a boy, a girl, a dog, a ball, a kite).

Give a direction and have a student drag the item to the correct spot.

Have the other students agree or disagree with the placement. If they disagree, have them tell or show where they think the item should be.

Put the dog under the tree.
Put the boy next to the house.
Put the kite above the house.
Put the ball under the swing set.

This can also be done using flat or solid shapes on the interactive whiteboard, having students drag the sphere above the cube or drag the hexagon next to the square.

Or:

Drag some items to places in the scene.

Have students use positional words to tell a partner where the item is located (e.g., the ball is in front of the house).

The Wing on a Flea

This book includes examples of triangles, rectangles, and circles and shows that the shapes can be many different items. The items in the story can be described using positional language.

Before Reading:

What shape do you think is the wing of a flea?
Let's read to find out.

During Reading:

Read *The Wing on a Flea: A Book About Shapes* by Ed Emberly.

As you read, discuss the shapes that are shown.

After Reading:

Show illustrations from the book to review positional language.

Discuss examples in the illustrations such as:

The goat is above the mountain.
The dragon is next to the mountain.
The pirate is behind the map.
The mouse is under the balloon.

Get Moving!

Students practice positional language on playground equipment.

Pose directions for students to stand in relation to items on the playground.

Stand above the slide.
Stand next to the swing.

Stand below the lookout tower.
Stand in front of the ladder.
Stand behind the pole.

Positional Language Mats

Partners place bears on a picture mat using positional language.

Materials: one privacy divider, two copies of a scene (such as a pond and trees or a playground), eight bear manipulatives (two of each color: yellow, red, blue, green)

Students play as partners.

Place the divider between the two students.

Each student gets the same picture and four bears.

Player A places the four bears on her picture.

Then she gives directions using positional language to Player B (e.g., place the green bear under the slide, put the yellow bear next to the trees, put the blue bear on the swing, and put the red bear above the basketball net).

Player B places his bears accordingly.

Once all bears are placed, the divider is removed and the two pictures are compared.

The goal is to have the bears in the same positions on both pictures.

Then the students switch roles and continue the activity.

Roll It

Students roll a cube to determine where to place a stuffed animal.

Each student has a stuffed animal.

A student rolls a cube to determine where each students' animal should be placed.

The teacher reads the position word that was rolled and then students move their stuffed animal to that position (e.g., if *beside* is rolled, the students put the animal beside them; if *over* is rolled, they put their animals over themselves).

The six sides of the cube can have any six positional terms (e.g., *behind*, *between*, *next to*, *in front of*, *above*, *below*, etc.).

Vocabulary

above	flat
behind	hexagon
beside	in front of
circle	next to
cone	rectangle
corner	shape
cube	side
cylinder	solid
face	sphere
figure	

ONLINE RESOURCES

General Resource

Additional Problems

Assorted Shape Cards

Introducing Geometric Shapes and Figures Student Observation Notes

Resources for Specific Activities

Playground Scene

Pond with Trees Scene

Using Positional Language Recording Sheet

These resources are available at http://hein.pub/MathinPractice, keycode MIPGK.

Students roll the cube to see where they should place their stuffed animal.

Your Partner Says

Partners practice positional language.

> Each pair has four figures: sphere, cube, cone, and cylinder.
>
> Player A gives a direction to position a solid figure:
>
> > *Put the cube behind your back.*
> > *Place the cone beside your arm.*
> > *Put the sphere in front of you.*
> > *Put the cylinder above your head.*
>
> Player B acts it out.
>
> Player A checks it.
>
> Partners switch roles and try it again.

MODULE 13

Exploring Geometry

About the Math K.G.4; K.G.5; K.G.6

Module 12 focused on students' introduction to geometric shapes, both flat and solid. Now, in Module 13, they further explore these shapes, analyzing and comparing them and describing ways in which they are alike and different.

The key ideas focused on in this module include:

- **analyzing and comparing two-dimensional and three-dimensional shapes through discussions and modeling**
- **combining simple shapes to form larger shapes through models and drawings.**

Once students have had experiences closely analyzing both two- and three-dimensional shapes, they revisit the similarities and differences between flat and solid shapes. Triangles and squares are alike because they are flat shapes. They are different because a triangle has three sides and a square has four sides. Or, students might notice that a cylinder and cone are alike because both have bases that are circles. Continued opportunities to compare shapes and discuss likenesses and differences deepen students' thinking about the shapes. (It is generally better to ask how shapes are *alike* or *different*, rather than *similar*, to avoid confusion later when students learn the specific meaning that *similar* has in geometry.)

Providing experiences in which students have to reason about shapes based on their attributes expands students' thinking. Students have seen and identified triangles, but they are challenged further when they see a three-sided shape that is not closed or a three-sided shape with one side that is not straight.

Nonexamples of a shape challenge students to think more deeply about the shape. They must take a stand and justify why it is or is not the shape for which they are searching (see Figure 13.1).

Figure 13.1 Students explain why the shapes in the circle are triangles and why the examples outside the circle are not triangles.

Creating models of shapes also focuses students on the attributes of the shape. By drawing or building shapes with straws and clay, Wikki Stix®, pipe cleaners, or blocks, students explore the attributes of the shape in a very personal and hands-on way. Using patterns or templates to draw shapes hones students' visualization skills as they observe and copy key elements of the shape.

Shapes can be used to form other shapes. Two triangles might form a square (see Figure 13.2). Like putting a puzzle together, students use trial and error as they move and turn shapes, putting them together to form larger shapes. They discover that by putting two squares side by side, they made a rectangle or that they could cover a yellow hexagon with six green triangles (pattern blocks). They are exercising and extending their spatial reasoning and observational skills and deepening their understanding of shapes.

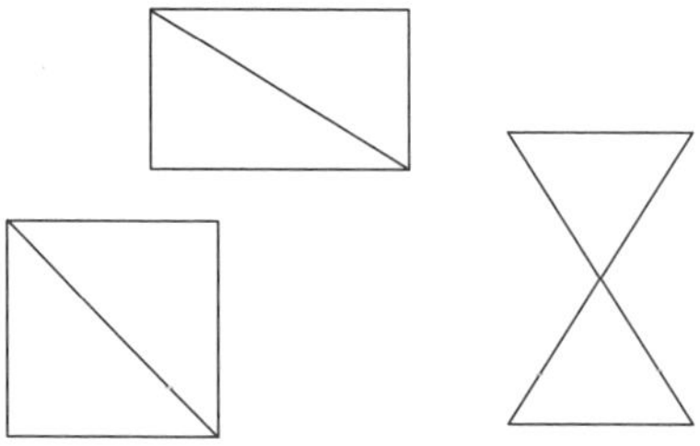

Figure 13.2 Students deepen their understanding of shapes as they combine them to form other shapes.

Construction activities with three-dimensional shapes also develop spatial reasoning as students build castles with variously shaped blocks, placing a cone on top for the spire (see Figure 13.3). They might use cubes to build larger cubes or rectangular prisms. They stack and build with the shapes and discover the ways the attributes of the figures (e.g., flat sides) allow the shapes to be put together to make something new.

Figure 13.3 This student stacks solid figures to create a castle as he explores three-dimensional shapes.

Exploring the Progression

PREVIOUS		NOW		NEXT
Preschool		**Grade K**		**Grade 1**
Possible experience identifying shapes in different sizes and exposure to making larger shapes from smaller ones with blocks and puzzle pieces	»	Analyzing, comparing, and describing flat and solid shapes; combining smaller shapes to form larger ones through drawing and modeling	»	Building or drawing shapes based on attributes; combining two- and three-dimensional shapes to form composite shapes; partitioning shapes into equal parts as a foundation for fractions

Learning Goals

I can compare flat shapes and solid figures.

I can make shapes from smaller shapes.

I can build with solid figures.

Lessons in This Module

Analyze and Compare Flat Shapes and Solid Figures

Sort Flats and Solids

Alike and Different

Triangle or Not a Triangle?

Creating Shapes

Draw Shapes from Riddles

Building Figures

Compose Shapes from Smaller Shapes

Mouse Shapes

Making Shapes from Other Shapes

Cover My Shape

Building Castles

Ideas for Instruction and Assessment

ANALYZE AND COMPARE FLAT SHAPES AND SOLID FIGURES

Students extend their understanding of geometric shapes/figures by exploring their attributes through comparisons.

Sort Flats and Solids

Note that sorting items in various ways is explored in Module 11.

Students sort a set of combined shapes (flat and solid) and talk about their thinking to better understand what it means to compare shapes.

Prepare sets of assorted solid figures and flat shapes for each student in the small group. In each set include several examples of two-dimensional shapes and also several examples of three-dimensional figures. The shapes can be in different sizes or orientations, but there should be only one of each specific shape (e.g., one square).

Give each pair a set of assorted shapes/figures.

Talk with your partner to decide on a way to sort the shapes.

Then, go ahead and sort them.

Be ready to tell us how you sorted them.

Students will likely sort the shapes by flat or solid, but may also sort them by size or even put squares with cubes, or circles with cylinders, to show that they recognize the connection between the flat shapes and the faces of the solid shapes.

Observe as students sort to identify what they are thinking about as they sort the shapes. If their first sorting idea doesn't work (e.g., sorting by shape), listen as they think of another way to sort them. Listen as they talk to see if they are using geometry vocabulary like *sides*, *faces*, *flat*, or *solid*.

Some students may initially want to sort by shape but soon find that there is only one of each shape. If they get stuck figuring out what to do next, ask them if they see ways that some of the shapes are alike. See if they can pull some shapes together that have something in common. If they are still having trouble, ask them if they see any 4-sided shapes (or another possible way of thinking about the shapes) to get them jump-started.

After students have sorted, ask:

How did you sort your items?

Did anyone think they could sort them by shape? Did you put all of the squares together or all of the circles together? Did that work?

So, what did you do? How did you find another way to sort them?

Allow students to share their thinking.

What did you think about? How were they alike?

So, you noticed some were alike in size so you sorted them as little and big? Did your group of little shapes have all different kinds of shapes?

That makes sense because any shape could be big or little.

Did anyone sort in a different way?

You sorted into flat and solid shapes? What shapes were in your flat group?

Begin a list of flat shapes on the board.

Did anyone else have a group that was flat?

Did you have any other shapes in your flat group?

Add other shapes to your list of flat shapes on the board.

What do we call shapes that are not flat? (Solid.)

What shapes were in your solid group?

Did anyone have different shapes in their solid group?

Begin a list of solid shapes on the board.

Did anyone sort them in a different way?

Allow students to share other possibilities if there are any.

Place a square and cube together and a circle, cone, and cylinder together.

Here are my shapes. What do you think I was thinking about when I sorted these? Tell your partner.

How are a square and cube alike? What do a circle, cone, and cylinder have in common?

Allow students to closely investigate your shapes.

What do you think?

Is it true that a cube has a square face?

Is it true that a cone and cylinder have a circle face?

Be sure all students can check the faces of the solid shapes to confirm this idea.

So a square is like a cube because of the faces on the cube and a circle is like a cylinder and cone because of the faces on them?

But they are different, too. How is a square different from a cube? How is a circle different from a cone or cylinder?

Shapes can be alike in some ways and different in other ways.

When we compare shapes, we think about how they are alike and different.

When we look at two shapes to compare them, what could we think about? (How many sides or corners they have. *Or*, Whether they are flat or solid. *Or*, What their faces look like.)

Alike and Different

Students pick two shapes and compare them to tell how they are alike and different based on their attributes.

For small group, have sets of solid figures and several sets of shape cards on the table.

Choose two shapes or figures that are alike in some way.

Be ready to tell how they are alike.

Observe as students select their items.

Turn and talk with your neighbor. Tell each other how your items are alike.

After they have talked to partners, have students share their ideas with the group.

Luke chose a rectangle and a square. How are they alike?

Can anyone think of another way his shapes are alike?

Anna chose a cube and a square. How are they alike?

Have other students share how their shapes and/or figures are alike.

Look at your shapes again. How are they different?

Turn and tell your partner one way that your shapes are different.

After they have talked to partners, have students share their ideas with the group.

How are Luke's rectangle and square different?

How are Anna's cube and square different?

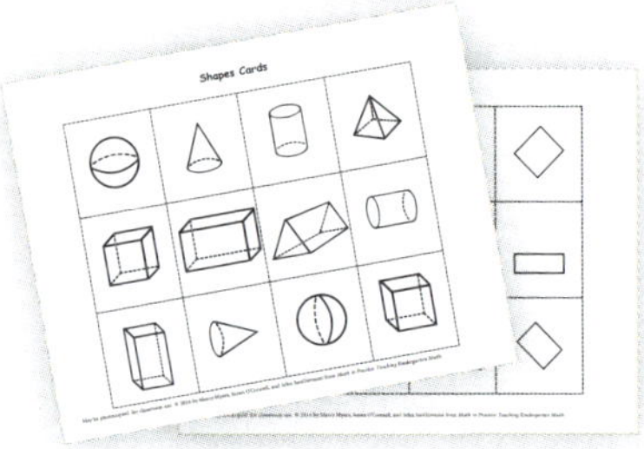

Then have students pick another set of shapes and/or figures.

Think about the two that you chose.

How are they alike?

How are they different?

Turn and tell a partner how they are alike and different.

Have students choose two cutout pictures of flat and solid shapes to glue into their math journals.

Have them write or tell how they are alike and different.

FORMATIVE ASSESSMENT

Show the student a cone and cylinder. Ask:

How are these figures alike? How are they different?

Students may say that they are both solid figures with curved sides and a face that is a circle. Or, that they are different because the cone only has one circle face or the cone has a point.

If a student is not able to express how the figures are alike and different, show the student one at a time and say:

Tell me what you know about this figure.

It may be easier for a student to tell about one figure rather than compare them, but once they have gained a better understanding of the figures, try comparisons again.

Record observational notes of student responses.

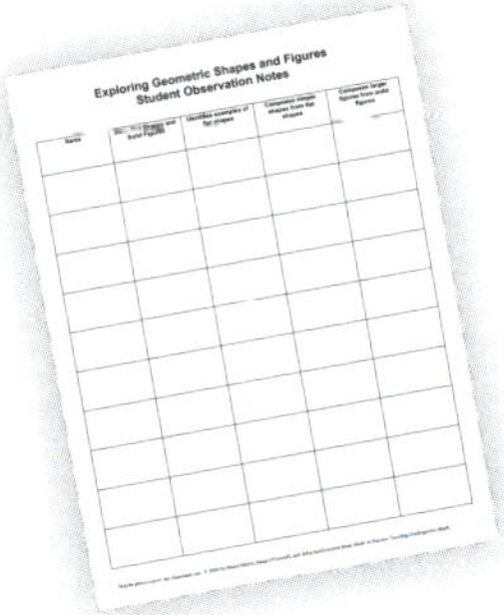

Use the observational notes page.

Show student a square and a triangle. Ask:

How are these shapes alike? How are they different?

Students may say that they are both flat shapes with straight sides. Or, that they are different because the triangle has three sides and the square has four sides (or three corners and four corners).

Triangle or Not a Triangle?

Students are introduced to the idea of examples and nonexamples of shapes and are asked to justify whether something is an example or not an example of a particular shape.

Each pair of students needs a set of cards that show examples and nonexamples of shapes. It may be helpful to also show the shapes on an interactive whiteboard or draw them on chart paper as the lesson progresses.

Give students the following shape card and show the shape on the board.

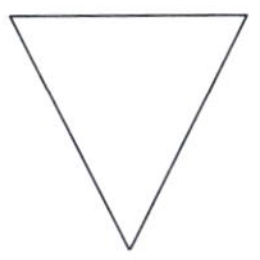

> *Even though you are showing the shapes on the board, giving students cards to hold allows them to take a closer look at each shape so they can closely examine them.*

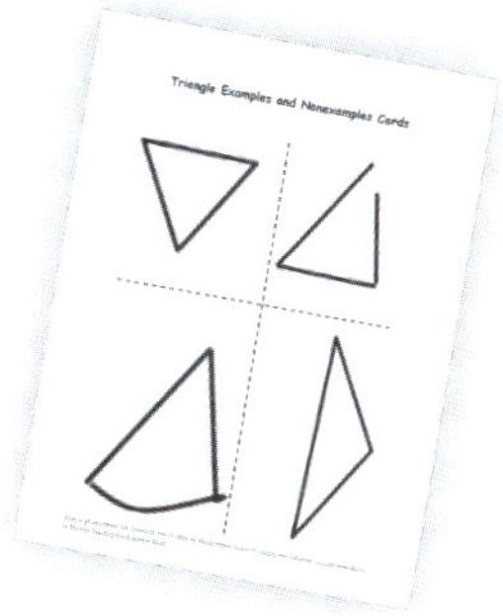

Turn and tell your partner the name of this shape.

What do we know about triangles?

Look at your card. Is it a triangle?

How do you know?

Does everyone agree?

Give students the following card and add this image to the board:

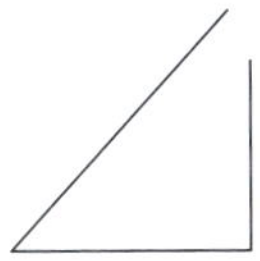

Talk to your partner: Is this a triangle?

Give students a minute to talk about it.

Does it look like a triangle?

What is different?

Does it have three corners? Count the corners.

Circle or highlight the corners, showing just two corners because the other one is *open*.

It does not have three corners so it is not a triangle.

On the board, cross out that figure.

Give students the following card and add this image to the board:

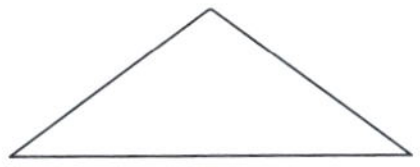

How about this shape?

What shape is this?

How do you know?

Are all of the sides touching to form corners?

Circle or highlight the corners, showing three of them.

This is a triangle.

Give students the following card and add this image to the board:

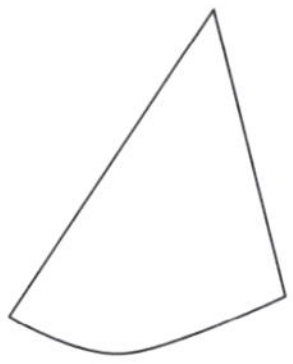

What about this one? Is it a triangle?

What do you notice about it? Does something look different?

Do triangles have curved sides?

Is it a triangle?

It is not *a triangle because one of its sides is curved, not straight.*

On the board, cross out that figure.

Let's add another shape that is a triangle. Show me one on your dry erase board.

Have students show their triangles.

What makes them a triangle?

Draw another triangle on the board.

Let's add another shape that is not *a triangle. Show me a shape that is* not *a triangle on your dry erase board.*

Have students show their nontriangles. Have them tell why their shape is not a triangle.

On the board, draw a 3-sided shape that is not a triangle.

Why is this not a triangle?

This part of the lesson can be saved for the following day to allow you to revisit the ideas with different examples.

Cross out the shape on the board.

Some of the shapes on the board are triangles and some are not examples of triangles.

We crossed out the shapes that are not examples of triangles. Sometimes people say they are nonexamples of triangles.

Continue with other examples using rectangles.

These can be introduced one at a time as was done with the triangles or all four cards can be given to partners to analyze together and then share their thinking with the small group.

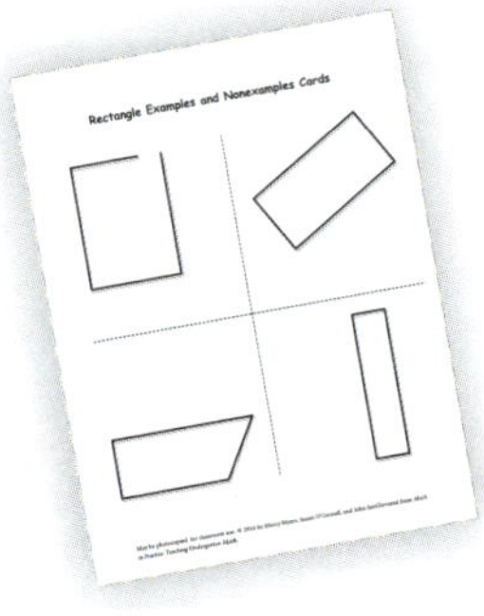

What do we know about rectangles?

Turn and talk with your neighbor about these shapes. Which ones are rectangles?

Listen as students discuss their ideas.

Discussions about the nonexamples should focus on the look of the corners and sides in the first figure and the sides not forming a corner in the fourth figure.

Continue with other shapes, showing an example and nonexample and asking students to justify why the shape is not a circle or hexagon or square.

Students at this level are not expected to know right angles or parallel sides, but they should recognize the look of the sides and corners/vertices of rectangles and recognize when a shape does not have that type of corner.

FORMATIVE ASSESSMENT

Students examine drawings of examples and nonexamples of shapes.

They are asked to circle examples of the shape and put an *X* on nonexamples.

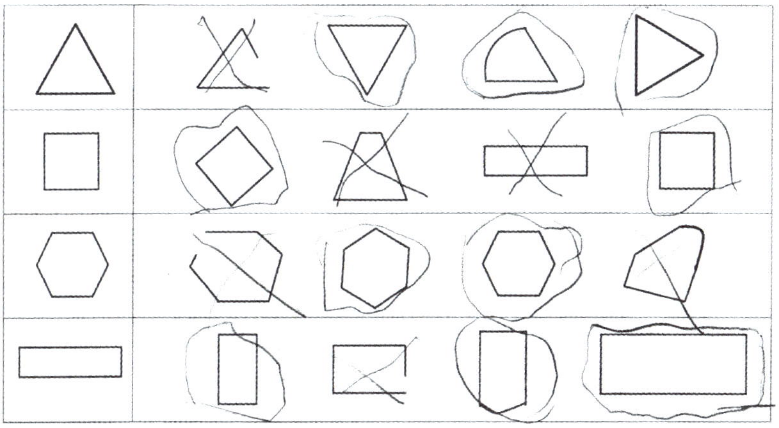

This student identifies shapes in traditional and turned orientations. He demonstrates an understanding that shapes do not have open sides by putting an *X* on those examples. He circled the triangle and rectangle drawings with a curved side.

Ask the student if the sides are all straight. Have him show you a side that is not straight. He may not recognize the difference between straight and curved sides. If he does recognize that some of the lines are curved, he needs clarification that triangles and rectangles do not have curved sides.

Creating Shapes

SHOW IT!

Students use their understanding of attributes of shapes to create flat shapes using clay, Wikki Stix, or pipe cleaners during center time and then bring their shapes to small group to share.

Have a basket with materials like clay, rollers, Wikki Stix, and/or pipe cleaners.

Show how you can make a square, circle, rectangle, triangle, or hexagon.

Have students create shapes and put them on a cookie sheet to bring to the small group.

In a small group, have them share their shapes.

What shapes did you make?
How did you make them?
Sarah made a square. How did you get the sides to be the same length?
Gavin made a circle with clay. I notice that he made a sphere first and then flattened it with his hand to make a circle.
Turn your shapes.
Are they still the same shape?
How do you know?

Differentiation

Some students may benefit from having a shape card to refer to as they work to recall the attributes of the shape.

Students work to create shapes using pipe cleaners. They can share and compare the shapes they made and show them in different orientations.

Draw Shapes from Riddles

Students draw shapes to strengthen their abilities to identify and visualize the characteristics of flat shapes.

Students use paper and pencils, crayons, or markers and dry erase boards to practice drawing shapes. It is helpful to have the Class Shape chart on the board for students to use as a reference.

I have some riddles that we are going to use to draw shapes.

While I'm reading, think about which shape it could be. Then draw it.

Pose the riddle:

This is a shape with three sides and three corners. What shape is it?

Watch as students draw a triangle.

What shape did you draw? What clues told you to draw a triangle?

Have students share their drawings with the class by holding up their paper or dry erase board.

Then pose another riddle:

This shape has four sides that are all the same length. What shape is it?

Watch as students draw.

What shape did you draw? What clues told you to draw a square?

Have students share their drawings with the class.

Continue with more riddles:

This shape has no sides and no corners. What shape is it?

This shape has four sides and four corners, but the sides aren't the same length. What shape is it?

This shape has six sides and six corners. What shape is it?

Building Figures

Students build solid figures from clay.

For small groups, have an assortment of different materials available, such as clay, miniature marshmallows, gumdrops, coffee stirrers, toothpicks, and straws. Have some solid figures available for students to use as a visual reference.

Start with something that might be easy.

Today we are going to build solid figures using some of these materials.

What do you know about a sphere?

Show how you can make a sphere.

Allow students time to work and then compare students' spheres.

Differentiation

Some students may need patterns or stencils to use if they have difficulty with fine-motor skills.

The Class Shape chart appears in Module 12.

Common Error

Triangles should include three straight sides and three corners, although some kindergarteners have trouble drawing straight lines, which is developmentally appropriate at this level. If you see any open shapes, help the student connect the sides to make a corner.

Drawing a hexagon could be difficult for most students. For this shape, pose the riddle and then model drawing a hexagon with the class or provide patterns for students.

Look at these two spheres.

One is blue and larger and one is red and smaller.

Are they both spheres?

How do you know?

Next, have students use materials to make a cube.

Show how you can make a cube.

Use any of the materials to make a cube.

Using clay is likely the easiest way for students to make solid shapes because they can roll it into cylinders or spheres or press it into cubes or cones.

Watch as students work.

Continue with cones and cylinders. For those figures that have circles as bases, students may use clay for the faces and use sticks to create the rest of the figure.

Additional Ideas for Support and Practice

The following ideas extend students' understanding of analyzing and comparing flat and solid figures and provide meaningful practice.

This activity takes just a few minutes and is a great way to revisit shape names and attributes throughout the year.

Pick One

Students do a quick comparison of two shapes that are displayed on a screen or under a document camera.

Display two shapes on the interactive whiteboard or under a document camera.

Have students compare the shapes and turn and tell their partners one way they are alike and one way they are different.

Shapes can be flat, solid, or one of each.

Variation: Display two shapes.

Ask questions and have students turn and tell their partners the answers.

Which shape has more (or fewer) sides?

Which shape has more (or fewer) faces?

Which shape has faces that are circles?

Which shape is solid?

Which shape rolls?

Talk About It/Write About It

Students draw pictures to define a shape.

Fold a piece of paper into fourths for each student.

Tell the name of a shape (e.g., triangle, circle, rectangle).

In the sections, students should do the following:

1. Draw the shape.
2. Write the shape name.
3. Draw two other examples of the shape.
4. Draw a shape that is a nonexample of that shape.

Grab a Shape

Students pick shapes and identify similarities with a partner.

Have a bag of assorted shapes.

Students pick a shape from the bag.

Students compare their shape with a partner to find a way they go together (are alike).

Students draw the two shapes and write (or tell) how they are alike.

Then, they replace the shapes and pick again.

COMPOSE SHAPES FROM SMALLER SHAPES

Students put shapes together to create larger shapes or figures, practicing their spatial reasoning as they move and rotate the shapes to make them fit together.

Mouse Shapes

Students listen to a story about how mice create pictures using shapes and then create their own pictures from shapes.

Students need various sizes and colors of paper squares, circles, triangles, rectangles, and hexagons to create pictures with the shapes.

Before Reading:

Review the names of the flat shapes.

Tell the students that you are going to read a story about how mice use shapes.

During Reading:

Read *Mouse Shapes* by Ellen Stoll Walsh.

As you read, discuss the shapes that the mice use to create other items.

After Reading:

Talk about what the mice created and have students name the shapes used for each.

It's important to remind students to arrange the smaller shapes and then glue them down.

Model a simple shape from the story using a document camera or easel.

Tell the students that they will be making their own pictures using shapes.

Brainstorm a list of ideas on the board. Suggestions may include a rocket, house, robot, train, car, or cat.

Take observational notes of the language students use to describe their work.

As students work, ask them to describe how they are moving the shapes to create their pictures.

When students have finished their pictures, have them talk with a partner and name the shapes they used.

Thinking Through a Lesson

Making Shapes from Other Shapes

Students use shapes to create composite figures.

This lesson works well in a small-group setting so you can listen to students' conversations and watch them as they compose shapes.

Have sets of thin attribute blocks or paper shapes for students to use to make composite shapes.

Did you know that shapes can be used like puzzle pieces?

Let's start with two squares.

SMP4

Give each student two squares from a set of thin attribute blocks (or paper squares that you have cut out).

How can you put these two squares together to make a new shape?

Give students time to try.

What shape did you make? (A rectangle.)

SMP3

How do you know it is a rectangle? (It has four sides and four corners.)

Look how Carey put her two squares side by side to make a rectangle.

Model for students who need help to visualize how to put the squares together.

Is there another way to make a rectangle using two squares? (We could put them going up and down.)

Jim put his squares together a different way to make a rectangle.

Creating shapes from other shapes provides a good review of vocabulary as students name the shapes they use and make.

Have students take two rectangles.

Put your two rectangles together to make a shape. What can you make?

Displaying other examples shows students that there's not just one correct way to place shapes.

Observe as students work.

Let's share what you made from two rectangles.

Tell us what you made and the shapes you used. (I made a really long rectangle with my two rectangles. *Or*, I made a square with my two rectangles.)

Common Error

Students may have difficulty with spatial reasoning as they work to compose new shapes. Have them work with partners so they can observe how others turn and flip the shapes. If needed, place your hand over their hands and manipulate a shape so they can feel the movement. Then allow them time to explore more on their own.

Encourage students to use shape names to practice vocabulary.

Have students pick a square and a triangle.

Make a new shape with your square and triangle. It's okay if it is not a shape you recognize or know a name for. Make any new shape you'd like.

Have students share some of their shapes.

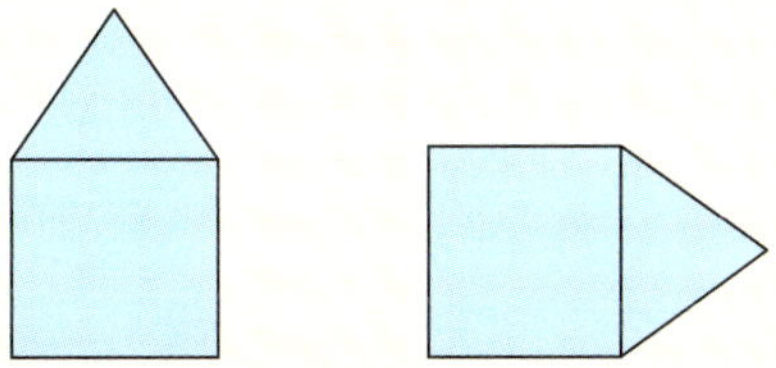

What shapes did you use to make your new shape? (A square and triangle.)

Have students pick any two shapes they'd like and create a new shape.

Be ready to tell us the shapes you used to make your new shape.

Letting students pick their own shapes and create imaginative composite shapes allows them to be creative and engages them in the task.

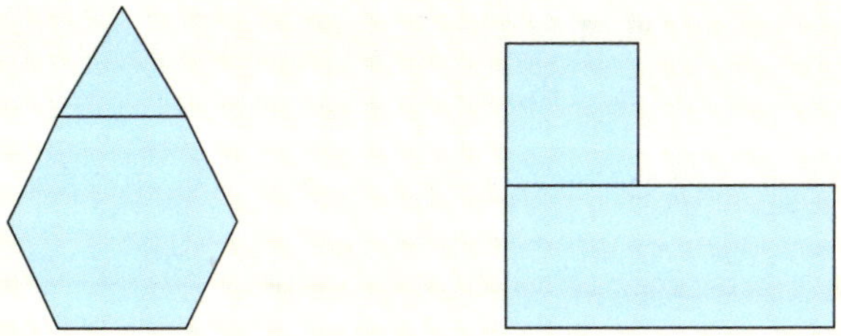

Have students share their shapes, naming the shapes they used to create their new shape.

As students are working with partners, invite a couple students who may still be struggling to work with you so you can observe them or offer tips to help them discover ways to manipulate the shapes.

Have students work with partners.

Use any two shapes to make a new shape.

When you've made your shape, turn and talk with your partner.

Tell your partner the shapes you used.

Allow students to try it several times.

Students then draw or trace their shapes in their math journal and write the names of the two shapes they used to make it.

Differentiation

For students who are not able to write the shape names, have labels available for them to glue into their journals.

Cover My Shape

Students practice manipulating shapes to cover a composite shape.

Prior to the lesson, trace pattern blocks on index cards to create composite shapes.

Show students the triangle, square, and hexagon pattern blocks and have them tell you the name of each shape as you show them the block.

Pattern blocks are not actually flat shapes, but are called by the names of flat shapes. Acknowledge this to students. Say that when they named these shapes they were looking at the faces and ask them to tell you what they see when they look at the faces. You can even laugh with students that someone must have made a mistake when they called it a square because it really isn't flat!

Ask them to justify how they know it is that shape.

Show students the trapezoid.

What can you tell me about this shape? How many sides and corners does it have?

Is it a rectangle?

Is it a square?

In what ways does it look different than a square or rectangle?

This is a trapezoid. *It has four sides and four corners, but it is a different shape than a rectangle or square.*

Students do not have to know a trapezoid at this level, but because pattern blocks are such common manipulatives in primary classrooms, students are generally exposed to the name/shape early. They explore the characteristics of a trapezoid in more detail in later grades.

Put some bowls of assorted pattern blocks around the table.

Do you remember when we put shapes together to make new shapes?

Choose two pattern blocks and make a new shape for me.

Give students time to make their shapes and then have them share them with partners, telling the names of the shapes they used.

Tell students that today you have some puzzles for them to solve.

I made some new shapes by combining two shapes and I need you to figure out which shapes I used to make my new shape.

I traced the shapes on cards.

You will need to look at what I traced and see if you can find the two shapes I used.

Then, cover them on your card with the blocks you think I used.

Look for students who can identify the pattern blocks by the traced image. Do they immediately grab for a block that matches?

Give each student an index card with a composite shape.

Watch as they try to find the blocks that fit on the traced shapes.

Observe as students place the blocks on the cards.

Have them tell you what shapes they used to cover the composite shape, then give them a new card.

After the lesson, place the shapes in the center area for students to continue practice.

This student covers shapes to figure out which two shapes were used to create the composite shape.

Common Error

Some students have a difficult time flipping or turning the blocks to get them lined up with the shape on the card. Suggest that they move the shapes in different ways to see which way might help it fit. Allow them to work with or watch partners. When suggestions or observations don't help, try placing your hand over theirs and moving the block to allow them to feel the movement. Then, let them explore again on their own.

Differentiation

To challenge students who find this task too easy, you might (1) make composite shapes that are formed with three or four blocks or (2) trace two blocks to form a composite shape and then erase the line between the shapes to make it more difficult for them to visualize the blocks that were used.

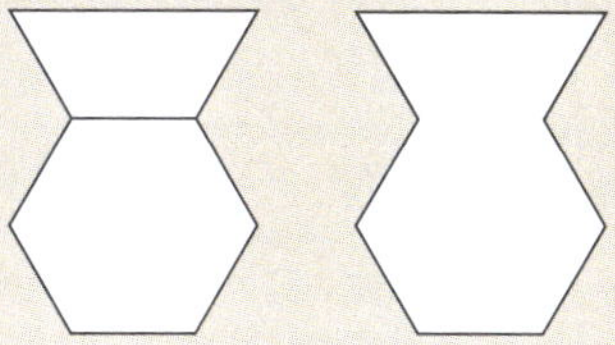

Differentiation

If several students seem to be moving through the task easily, allow them to work together and tell each other the shapes they used after completing each card. Then, spend extra time with the students who need support to find the blocks and cover the shapes.

FORMATIVE ASSESSMENT

Lay out some pattern blocks or attribute blocks and say:

> *Can you make a new shape with two squares? Show me your new shape. Tell me about your shape.*
>
> *Can you make a new shape with two triangles? Show me your new shape. Tell me about your shape.*

Record observational notes of the student's work and responses.

Students may make a new shape they can name (e.g., a rectangle or square) or may make a new shape that they describe as looking like something (e.g., a kite or an hourglass).

Use the observational notes page.

Exploring Geometric Shapes and Figures
Student Observation Notes

Building Castles

Students observe a castle made of solid figures and try to make another like it.

Prepare several sets of solid figures for students to use during small-group instruction.

> *I am going to build a castle with solid figures.*
>
> *Watch as I do it and then I am going to see if you can make one just like it.*

As students watch, put two large cubes side by side and a cylinder on top of each cube.

> *What shapes have I used so far?*
>
> *I want my castle to be taller.*

Put a cone on top of each cylinder.

> *What did I add to my castle?*
>
> *Use the figures in front of you.*
>
> *Can you make a castle like mine?*

Allow students to build independently.

As students work, notice how they position the figures. Do they start with the cubes and then stack the cylinders on top? Do they understand how to rotate or turn the figures as necessary? Can they adjust their thinking if their figure doesn't look the same (trial-and-error strategies)?

When students have finished building, talk about the figures they used and any trial-and-error moves that you noticed.

> *I notice that Brady used a large cylinder on the right and a smaller cylinder on the left.*
>
> *What happened when you tried to put a large cone on top of that little cylinder?*
>
> *What did you do?*
>
> *Did it work?*

Then, have students use any six figures they'd like to build their own castle.

Note the connection to counting. Depending on the number of materials available, this could be modified to more or fewer figures used to build each castle.

As they work, notice how they position the figures and pose questions to help guide their thinking.

What figures are you going to start with?
What could you put on top of your cylinder?

When the castles are built, ask questions to prompt thinking and comparisons.

Did anyone use a sphere as they were building? Why or why not?
How could you include a sphere?
Turn to a partner: Which of you used more cubes to make your castle?
Look at these two castles. How are they alike? How are they different?

Additional Ideas for Support and Practice

The following ideas extend students' understanding of composing shapes and provide meaningful practice.

Composite Pictures

Have students create pictures (e.g., a house, an animal) using attribute blocks.

When the picture is complete, students count the number of each shape they used to make their picture and record it in their math journal (e.g., four squares, no triangles, one rectangle).

Take pictures of the students' creations and project them for the class to see.

Have students tell about their picture and report the number of each shape they used.

Talk About It/Write About It

Build a tower with a cone, a cube, and a cylinder.

Can you build it a different way?

Which figure can't be on the bottom? Why?

Pattern Block Pictures with Partners

Students work with a partner to create pictures using pattern blocks.

Each pair of students needs a bowl of pattern blocks.

Player A uses pattern blocks to create a simple picture.

Player B uses pattern blocks to create a duplicate of Player A's picture.

They compare to be sure their pictures are the same.

Then Player B creates a simple picture and Player A copies it.

It is best to limit the number of pattern blocks used in each picture (e.g., eight or fewer), so players are not waiting a long time for their partner to make a picture.

Vocabulary

alike	figure
build	flat
circle	hexagon
compare	rectangle
cone	shape
corners	side
cube	solid
cylinder	sphere
different	square
draw	triangle
face	

ONLINE RESOURCES

General Resources

Additional Problems

Exploring Geometric Shapes and Figures Student Observation Notes

Resources for Specific Activities

Examples or Nonexamples Recording Sheet

Rectangle Examples and Nonexamples Cards

Shapes Cards

Triangle Examples and Nonexamples Cards

These resources are available at http://hein.pub/MathinPractice, keycode MIPGK.

Tangram Puzzles

Students explore tangram puzzles to make composite shapes/pictures.

Students use a set of tangram shapes (seven pieces are in one tangram).

Pose a task:

How many squares can you make with the shapes in your tangram?

How many triangles can you make?

Can you make a house? How many shapes did you use? Can you do it another way?

Can you make a snake? What other animal can you make?

Building Castles with Partners

Students work with a partner to build and copy castles. This center activity is done following the lesson Building Castles.

Pairs of students need sets of solid figures.

Player A starts by building a castle using several solid figures.

Player B then copies Player A's castle, and they compare structures.

Then Player B builds a castle and Player A copies it.

Pattern Block Letters

Students use pattern blocks to create letters.

Put a bowl of pattern blocks in the center area and a strip showing capital letters.

Have students use the pattern blocks to create letters.

They can choose any letters they'd like and use any blocks they'd like, but some letters are easier to form (e.g., *F, E, H, I, L, T*), and others offer a challenge (*B, C, D, G, O*).

Challenge: Have them form the letters in their first name.

Recommended Children's Literature

Baker, Keith. 2004. *Quack and Count*. Boston: HMH Books for Young Readers.

Calmenson, Stephanie. 1995. *Dinner at the Panda Palace*. New York: HarperCollins.

Carle, Eric. 1996. *1, 2, 3 to the Zoo*. New York: Philomel Books.

——. 1999. *Rooster's Off to See the World*. New York: Alladin.

Christelow, Eileen. 1993. *Five Little Monkeys Sitting in the Tree*. Boston: HMH Books for Young Readers.

——. 2012. *Five Little Monkeys Jumping on the Bed*. Boston: HMH Books for Young Readers.

Crews, Donald. 1995. *Ten Black Dots*. New York: Greenwillow Books.

Cronin, Doreen. 2006. *Click, Clack, Splash*. New York: Atheneum Books for Young Readers.

Ehlert, Lois. 1992. *Fish Eyes: A Book You Can Count On*. Boston: HMH Books for Young Readers.

Emberly, Ed. 2015. *The Wing on a Flea: A Book About Shapes*. Los Angeles: AMMO Books.

Fleming, Denise. 1995. *Count!* New York: Square Fish.

Florian, Douglas. 2000. *A Pig Is Big*. New York: Greenwillow Books.

Franco, Betsy. 2009. *Zero Is the Leaves on the Tree*. New York: Tricycle Press.

Giganti, Jr., Paul. 1994. *How Many Snails? A Counting Book*. New York: Greenwillow Books.

Hague, Kathleen. 2015. *Ten Little Bears: A Counting Rhyme*. Seattle, WA: Two Lions.

Lee, Mark. 2013. *Twenty Big Trucks in the Middle of the Street*. Somerville, MA: Candlewick.

Lottridge, Celia. 2012. *One Watermelon Seed*. Markham, Ontario, Canada: Fitzhenry & Whiteside.

Marshall, James. 1998. *Goldilocks and the Three Bears*. New York: Puffin Books.

Medearis, Angela Shelf. 1996. *The 100th Day of School*. New York: Cartwheel.

Murphy, Stuart J. 1972. *The Best Bug Parade*. New York: HarperCollins.

——. 1997. *Just Enough Carrots*. New York: HarperCollins.

——. 2000. *Monster Musical Chairs*. New York: HarperCollins.

Otoshi, Kathryn. 2010. *Zero*. Mill Valley, CA: KO Kids Books.

Reid, Margarette S. 1995. *The Button Box*. New York: Puffin Books.

Savage, Stephen. 2013. *Ten Orange Pumpkins*. New York: Dial Books.

Thong, Roseanne. 2014. *Round Is a Mooncake: A Book of Shapes*. San Francisco: Chronicle Books.

Tompert, Ann. 1996. *Just a Little Bit*. Boston: HMH Books for Young Readers.

Walsh, Ellen Stoll. 1995. *Mouse Count*. Boston: HMH Books for Young Readers.

——. 2007. *Mouse Shapes*. Boston: HMH Books for Young Readers.

——. 2010. *Balancing Act*. New York: Beach Lane Books.

Wise, William. 2004. *Ten Sly Piranhas*. New York: Puffin Books.

Wood, Audrey. 2004. *Ten Little Fish*. New York: Blue Sky Press.

Yektai, Niki. 2001. *Bears at the Beach*. Brookfield, CT: Millbrook Press.